BRAIN-FRIENDLY LEADER

The Human Edge In The Age Of AI

SILVIA DAMIANO
JUAN CARLOS CUBEIRO
RELMI DAMIANO

about my brain®

“

THE FUTURE WILL NOT BE DIVIDED BETWEEN PEOPLE WHO USE AI AND PEOPLE WHO DON’T. IT WILL BE DIVIDED BETWEEN PEOPLE WHO CAN STILL THINK AND PEOPLE WHO HAVE FORGOTTEN HOW.

”

ENDORSEMENT

BY JOHAN BARTON

SENIOR LEADERSHIP CONSULTANT
WELLINGTON, NEW ZEALAND

The core message of Brain-Friendly Leader: The Human Edge in the Age of AI, the latest from the gifted Silvia Damiano, together with Juan Carlos Cubeiro and Relmi Damiano, is indisputable. Its content is compelling, deeply researched and shaped by the authors' combined experience across leadership, human development, culture, creativity and transformation.

The authors quickly bring the reader to a clear understanding of why traditional behavioural approaches to leadership development are no longer sufficient in a world being reshaped, at speed, by artificial intelligence. They take the reader beneath the visible behaviours of leadership, past strategy, past technique, and into the neuroscience, mindset and inner condition that shape how leaders actually think, feel, decide and act.

Silvia's talent for making complex neuroscience simple and accessible is beautifully strengthened by Juan Carlos Cubeiro's deep expertise in leadership, talent and organisational transformation, and by Relmi Damiano's creative, human-centred lens. Together, they offer something rare: a genuinely integrated view of leadership, rather than three disciplines stitched together. The book's real achievement is showing that alignment, integration and self-regulation are not soft ideas, they are trainable, and they are what allow imagination and sound judgement to survive contact with pressure. Through the i4 Neuroleader Model, the authors give the reader a concrete way to build and sustain that alignment, rather than simply naming it as a goal.

I unreservedly endorse this book. It does more than diagnose the moment leaders find themselves in, it offers a credible, well-evidenced way through it, and I have no doubt it will become a go-to leadership approach for the years ahead. Read it. You will sustain The Human Edge in the Age of AI.

BRAIN-FRIENDLY LEADER
THE HUMAN EDGE IN THE AGE OF AI

A book by Silvia Damiano, Juan Carlos Cubeiro and Relmi Damiano

English Edition 1
Published on July 23rd 2026
By About my Brain Institute Pty. Ltd.

Email hello@aboutmybrain.com
Web www.aboutmybrain.com
ISBN 978-0-6459537-5-6

Cover & Design by Relmi Damiano
Editing by Silvia Damiano, Relmi Damiano
Manuscript Review and Foreword by Dr Courtney McBath
Manuscript Review and Endorsement by Johan Barton

6 5 4 3 2 1

CONTENTS

INTRODUCTION

PART 1: THE HUMAN SYSTEM BENEATH LEADERSHIP

PART 2: A WORLD MOVING FASTER THAN THE HUMAN SYSTEM

PART 3: THE FOUR HUMAN ABILITIES OF LEADERSHIP

PART 4: LEADING IN AN AGE OF ARTIFICIAL INTELLIGENCE

CONCLUSION

FOREWORD

BY DR COURTNEY MCBATH

EXECUTIVE COACH
VIRGINIA, USA

"There is a point, toward the end of any serious inquiry into leadership, where the question itself changes. It is no longer about what leaders should do. It becomes about what allows a human being to function well enough to lead at all." p. 121

Have you reached that point in your leadership journey where the question itself has changed?

If your experience is anything like mine, you have spent more time, money and energy than you ever imagined trying to become a better leader by changing your behavior. For years, my focus was fixed on doing better so I could become better. Then I became a practitioner of the i4 Neuroleader™ Model and began to understand something far more freeing: leadership is not only about better performance. It is about becoming a more integrated human being.

In this book, Silvia Damiano, Juan Carlos Cubeiro and Relmi Damiano invite you to escape the minimized life of "human doing" and rediscover the freedom of "human being."

They show you the conditions your brain needs to transform your ability to perform, collaborate, innovate and lead with agility. Instead of performing in the traditional sense, you are invited to integrate every part of who you are internally so your leadership can be expressed more powerfully externally.

Leaders often feel driven to perform quickly, perfectly and relentlessly. That pattern has too often become the source of burnout, poor health and deep personal frustration. In Brain-Friendly Leader, you will discover the power of aligning your mind, body and heart so your leadership flows from a healthy place within. This is a path from a dysregulated system to an aligned one.

Like so many of us, you may have lived under constant pressure to "just get it done." Yet human beings require consistent rhythms of rest, recovery and renewal. The authors write with scientific grounding, practical wisdom and human warmth. They do not preach about leadership. They reveal the conditions that allow a human being to lead well.

Effective leadership depends on our ability to connect those around us to an inspiring, shared vision. This book walks you through the conditions required to make collaboration systemic and sustainable. The power of collaborating with others through ongoing inspiration is desperately needed in today's world.

I am not talking about the ability to deliver an inspiring speech. I am talking about what the authors describe as "the quieter and more demanding capacity to generate energy, meaning and direction without coercion, and to transmit that state in a way that activates something in others."

When you are personally at peace, well rested and confident, you bring that presence into every space. People are not inspired by what you say as much as they are by who you are. Brain-Friendly Leader reminds us that leadership has less to do with possessing a rare gift and much more to do with activating the areas of the brain and body that leadership requires.

This book reawakens the imagination that pressure often buries. The authors point to practical ways to access the lost art of seeing beyond current limitations and exploring the possibilities of a future that does not yet exist.

All of your leadership success to date has been rooted, in some way, in your ability to innovate. You have produced ideas that solved problems, improved processes and added value to the lives of those around you. We have not lost the ability to imagine. It may simply have been crowded out by urgency, pressure and noise.

What if we used AI for the tasks that do not require our full human intelligence and then spent the time we saved thinking? Dreaming. Imagining. Growing.

This book will walk you through simple steps to flourish in optimism, resilience, curiosity and idea generation. You were born to do what artificial intelligence cannot do on its own: combine curiosity, imagination, meaning and lived experience to create what has never been.

Ultimately, the human leader has an agility that cannot be replaced by artificial means. Through intuition, we can make decisions and access intelligence that is not always immediately visible or encoded. Brain-Friendly Leader is not based on untested philosophy. It stands on years of research, testing, evolution and practical application.

That research suggests intuition is not mystical guesswork. It is a form of embodied intelligence, shaped by experience, pattern recognition and signals we often dismiss before we understand them. Within the first few pages, you will begin to resonate with this powerful idea: design, not luck, and biology, not chance, fuel the best leadership.

When you become brain-friendly, you become more knowledgeable about how your brain and body function. That knowledge can unlock potential you may never have known existed.

For too long, I allowed the tyranny of the urgent to co-opt the power of my brain. The very capacity I needed to dream, imagine and create new solutions was being consumed by pressure. Often, what I claimed were new ideas were simply old policies and procedures in better clothing.

But as I learn to live with more rest, recovery and brain stamina, I am learning to function well enough to lead, and to lead well. I am finally asking the right question:

"What allows a human being to function well enough to lead at all?"

As you pick up this book, prepare to be challenged not only in what you think, but in how you think. Take notes, answer the reflection questions and take real steps toward brain-friendly leadership.

Get ready to unlock your human intelligence edge in this daunting age of artificial intelligence.

ABOUT THE VOICES IN THIS BOOK

THIS BOOK BRINGS TOGETHER THREE COMPLEMENTARY PERSPECTIVES ON LEADERSHIP AND HUMAN DEVELOPMENT

Silvia Damiano, Founder of the About my Brain Institute, is the originator of the i4 Neuroleader™ Model and Methodology. She brings the scientific and methodological foundation of this work, a deep inquiry into the human system beneath leadership, and the lived experience from which the model first emerged.

Juan Carlos Cubeiro, one of the leading voices in leadership, talent, and transformation in the Spanish-speaking world, brings more than three decades of experience advising organisations, coaching senior executives, and writing extensively on management, human potential, and the future of work. He contributes the practical and humanistic breadth of the book, drawing from business, sport, philosophy, and the long history of human flourishing.

Relmi Damiano, Co-Founder of the About my Brain® Institute, brings the creative, relational, and generational perspective that has helped the model take form in the world. A visual thinker and designer, she represents a generation already living and working at the speed of artificial intelligence.

Full biographies appear at the end of this book.

Together, these three voices give this book its distinctive character: rigorous, relational, practical, and deeply human.

“

THE FUTURE OF LEADERSHIP WILL NOT BE DECIDED BY WHAT WE DO, BUT BY THE STATE WE DO IT FROM.

”

FROM HABITS TO WHAT LIES BENEATH

Most leaders know the feeling. You know what the moment requires, and still something in you cannot quite meet it. You want to stay calm, but pressure gets there first. You want to listen, but urgency narrows the room. You want to lead from clarity, and find that clarity has already left the building.

Years ago, this gap was the reason I made Make Me a Leader, a documentary that took me across continents to sit with leaders in the middle of exactly this struggle — capable, well-intentioned people who could describe the leader they wanted to be with total clarity, and still watched themselves fall short of it under pressure. What struck me, filming conversation after conversation, was that almost no one blamed a lack of knowledge or will. They blamed something closer to wiring. A gap between intention and behaviour that seemed to open precisely when the stakes were highest.

In our previous book, Brain-Friendly Leadership: The 9 Habits of a Future-Fit Mind, we began with a conviction that has only strengthened over time. Leadership cannot be separated from the brain. Every thought we think, every emotion we feel, every decision we make, and every action we take is shaped by the most complex organ we possess. If leadership is expressed through human behaviour, it is inevitably influenced by the biological system from which that behaviour arises.

That first book emerged from a desire to make this understanding practical. Much of the language surrounding leadership remains abstract, often detached from the lived reality of those who must lead in complex environments.

We wanted to offer something different: a framework people could apply, experience, and integrate into daily life. The result was a set of nine habits, organised across direction, care and development, designed to support clearer thinking, more intentional self-care, and the mental and emotional conditions required to navigate an increasingly demanding world.

Those habits mattered, and they continue to matter. They gave many leaders a language for change and a structure through which leadership could be understood as something developable rather than fixed. They were grounded in a fundamental insight from neuroscience: the brain is plastic.

It changes with experience, repetition and what we consistently reinforce. From this perspective, leadership is not reserved for a gifted few. It is shaped over time. And yet, as our work continued across organisations, industries and cultures, another pattern began to emerge, one that could not be explained by knowledge or intention alone. People can know what to do and still find themselves unable to do it.

They can understand the importance of presence and still become reactive. They can value empathy and still lose access to it under pressure. They can cultivate purpose, adopt supportive routines and engage with powerful models, yet revert to familiar patterns precisely when the stakes are highest. Practising a habit in stable conditions is not the same as sustaining it when pressure rises, uncertainty intensifies and the human system begins to strain.

Most leaders recognise this gap, even if they do not always name it. It appears in moments that feel unremarkable on the surface yet prove decisive in their impact. A conversation meant to build trust becomes subtly guarded. A decision that required perspective becomes narrowed by urgency. A leader who values calm finds themselves transmitting pressure instead. The intention is still there, but access to it has shifted. This is not a failure of capability. It is a change in state.

At that point, a deeper question begins to surface. Not only how leadership can be developed through habits, but what makes those habits possible in the first place. Beneath the visible world of behaviour lies another layer, less obvious yet far more decisive. It is the layer of internal state, physiological regulation, emotional tone and continuous signalling between brain and body. It is here that a leader's capacity is either supported or compromised, long before a word is spoken or a decision is made.

For decades, leadership development has focused primarily on what can be observed. Behaviours, competencies and performance markers have provided a practical way to define and measure effectiveness. This is understandable. Behaviour is visible. It can be taught, evaluated and reinforced. But behaviour is not the beginning of leadership. It is its expression.

Before a leader communicates, a system is already in motion. Before a decision is made, the brain and body have already entered a particular state. The ability to remain composed, think clearly, listen deeply or act with courage depends not only on intention, but on whether the system can sustain those capacities in that moment.

This is the limitation of a purely behaviour-based approach. It assumes that once people understand the right principles and rehearse the right actions, those actions will hold. Yet human beings are not machines executing instructions. We are living systems, continuously influenced by stress, energy, emotion, environment and experience. A person may know exactly how they intend to lead and still be unable to access that version of themselves when their system is dysregulated. In such moments, behaviour is shaped less by intention than by pattern. This is why the foundation of our work has always needed to extend beyond behaviour.

At the centre of that foundation sits the i4 Neuroleader Model, created by Silvia to explain how human beings can strengthen four essential leadership competencies: performance, collaboration, innovation and agility. It emerged from four years of research at the intersection of neuroscience and leadership, shaped by formal inquiry, conversations with leaders across the world and Silvia's own lived engagement with the practices from which the work first emerged. From the beginning, Relmi was part of that evolution, helping translate the model into forms people could recognise, enter and experience for themselves.

Juan Carlos has brought to this lineage the perspective of someone who has spent more than three decades developing leaders and talent in real organisational systems across Europe and Latin America, and who knows that transformation becomes sustainable only when it begins beneath behaviour.

The wider model includes sixteen human abilities, four of which sit at its core: Integration, Inspiration, Imagination and Intuition. These four are not the whole model, but they are foundational to it. They underpin the competencies of performance, collaboration, innovation and agility, and offer a powerful way of understanding what makes leadership effective, sustainable and deeply human. We will return to these i4 abilities in greater depth in Part III of this book.

What matters at this stage is not to understand the model in full, but to recognise what it is pointing to. Leadership does not emerge from behaviour alone. It emerges from the condition of the human system.

When *Integration* is present, a leader can remain steady and coherent even as pressure increases. When it is not, fragmentation appears and behaviour begins to lose alignment.

When *Inspiration* is available, energy becomes directed and meaningful, and when it is absent, effort becomes mechanical or depleted.

When *Imagination* is active, new possibilities can be created. When it is constrained, thinking becomes repetitive.

When *Intuition* is accessible, complexity can be navigated with discernment. When it is muted, decision-making becomes either overly analytical or reactive.

These are not abstract ideas. They are human capacities that determine whether leadership can actually be expressed in real time. Their importance becomes even more apparent in the world we now inhabit.

We are living through a period of transformation so fast that it is still difficult to comprehend fully. Two years ago, none of the leaders we work with were asking us what to tell their children to study. Now almost all of them are. Conversations that once took place in boardrooms about strategy and performance now include a more personal question about what human beings will actually do with themselves. That question is not theoretical. It is showing up in kitchens, staff meetings, one-to-one coaching sessions and in the minds of people who have spent thirty years building expertise that may no longer mean what it once did.

The impact of this shift is not only economic. It is cognitive, emotional and existential. Artificial intelligence does not simply change what we do. It influences how we think. It alters our relationship with memory, attention, language and imagination. It can enhance efficiency while increasing dependency. It can expand capability while weakening internal anchoring.

Artificial intelligence is not only replacing tasks. It is replacing effort: the effort to remember, structure thought, sit with complexity and tolerate uncertainty long enough for insight to emerge. And while this creates extraordinary opportunity, it also introduces a subtle risk. As external intelligence expands, internal engagement begins to contract.

We may become more productive while becoming less reflective, more connected while less present. The deeper challenge of our time may not be technological advancement itself, but the speed at which it is unfolding relative to our capacity to remain integrated. We speak extensively about innovation, yet far less about what happens when innovation overlooks the human. We celebrate acceleration, yet rarely consider its effect on our nervous systems, our relationships, our children and our sense of meaning.

This is why the conversation about leadership must evolve.

It is no longer sufficient to ask what leaders do. We must ask from what state they do it. We must understand the system that produces behaviour, the level of coherence or fragmentation a leader carries into a moment, and how that shapes everything that follows. Because leadership is not only what we execute. It is what we transmit. And increasingly, it is what we co-create through the quality of our conversations. In our work across generations, this has become unmistakable. The ability to sit with another human being, to think together and to remain present without rushing to resolution is no longer a soft skill. It is a cognitive and relational necessity. It is also where much of this work comes alive, not in theory but in the lived experience of dialogue.

This book is an invitation to explore that deeper layer. It does not reject habits, nor does it dismiss skill or discipline. It reframes them. Habits remain important, but they are expressions of something more fundamental. Leadership begins with the system that makes behaviour possible.

In the chapters that follow, we examine that system in greater depth. We explore its bioelectric and neurobiological foundations, the ways stress and overload influence its functioning, and the four human abilities at the core of the i4 Neuroleader Model.

We also consider the broader context in which leadership now takes place, shaped by artificial intelligence, shifting definitions of work and an increasing need to preserve what remains fundamentally human.

Because ultimately, the question at the centre of this book is not only how to lead more effectively. It is how to remain sufficiently connected to our own humanity to lead wisely. And perhaps that is the real frontier. Not behaviour. Not performance. Not intelligence alone. But the capacity to remain connected to the living system within us, and to one another, while the world around us becomes faster, louder and more artificial. That is not the edge of leadership. It is its beginning.

PART I

THE HUMAN SYSTEM BENEATH LEADERSHIP

BEFORE LEADERSHIP BECOMES VISIBLE IN BEHAVIOUR, IT IS ALREADY UNFOLDING WITHIN THE HUMAN SYSTEM

CHAPTER 1

THE HUMAN SYSTEM
YOU ARE NOT JUST A BRAIN

For a long time, we have told ourselves a convenient story about leadership. It places thinking at the centre of everything, as if leadership were primarily a cognitive act: analysing situations, making decisions and choosing the right course of action. From this perspective, the brain becomes the central actor, a kind of command centre that receives information, processes it and directs the rest of the organism accordingly. It is an appealing idea, in part because it gives us a sense of control. If leadership is fundamentally about thinking, then improving leadership becomes a matter of improving thought.

This assumption has shaped how leadership is developed. We train leaders to think strategically, reframe challenges, apply models and sharpen their judgement. We invest heavily in cognitive capability, often assuming that clearer thinking will naturally produce better leadership.

And yet, when we observe leadership in real conditions, under pressure, in complexity, in uncertainty, something does not quite fit this view. Leaders do not operate as detached thinkers observing the world from a distance. They experience fatigue, emotional load, tension and competing demands. They react before they reflect. They sense shifts in a room before they can articulate them. They often find themselves acting in ways that do not fully correspond to what they know, believe or intend.

Most leaders recognise this immediately. The meeting where you intended to listen and found yourself interrupting. The conversation where you knew what mattered and still said the wrong thing. The moment you walked away thinking, “That is not how I wanted to show up.”

This is not an exception to leadership. It is part of its reality. Which leads to a more fundamental question: what if leadership cannot be fully explained by thinking alone? Or, more precisely, what if the brain, while essential, is not sufficient to explain how leadership actually unfolds?

The brain is part of the system, not the system itself. This shift is subtle, but once seen, it changes everything. It invites us to move away from a brain-centred understanding of leadership and toward a more integrated view of the human being. The brain does not operate in isolation. It exists in continuous interaction with the body and nervous system, forming a dynamic network that shapes how we perceive, interpret and respond to the world.

Over the past decades, advances in neuroscience and physiology have increasingly challenged the idea that cognition stands apart from the rest of the organism. Antonio Damasio's work, for example, showed that decision-making is inseparable from emotion, and that emotion itself is deeply rooted in bodily processes. Individuals with impaired emotional processing could still reason logically, yet struggled to make even simple decisions, revealing that cognition alone is not sufficient to guide action (Damasio, 1994). What appears to be rational choice is, in fact, supported by signals arising from within the body.

A similar line of inquiry emerges in research on interoception, our capacity to sense the internal condition of the body. As neuroscientist A. D. Craig (2002) has shown, the brain is continuously receiving and interpreting signals from within the body, using them to shape how we experience emotion, perception and self-awareness. What we experience as thought, feeling or judgement is therefore influenced by physiological conditions that operate largely outside conscious awareness. Thinking does not occur in a vacuum. It is shaped by the condition of the system in which it arises.

When leadership is viewed through this lens, the limitations of a brain-only model become more apparent. Leaders are not disembodied thinkers. They are living systems, shaped by the ongoing interaction between brain, body and environment. Their capacity to think clearly, decide effectively and act with intention is inseparable from the state of that system.

At any given moment, the human organism is engaged in a continuous process of regulation. The nervous system scans for cues of safety or threat, adjusts levels of arousal, directs attention and prepares responses. Much of this activity takes place automatically, outside conscious awareness. Before a leader has formulated a response, the system has already begun to shape what is perceived as relevant, what is felt as urgent and what is experienced as possible.

Stephen Porges' work on the autonomic nervous system offers one way of understanding this dynamic. His Polyvagal Theory suggests that physiological state influences whether we are oriented toward connection and openness or toward protection and defence. When the system detects safety, it supports social engagement, flexible thinking and emotional regulation. When it detects threat, it narrows attention, increases reactivity and prioritises survival (Porges, 2011). This shift is not deliberate. It is biological, and it has direct consequences for how leadership is expressed. The ability to remain present, listen deeply, think clearly and respond rather than react depends on whether the system is in a state that allows those capacities to emerge.

A leader may understand the importance of empathy, yet find themselves unable to access it when their system is overwhelmed. They may value composure, yet experience a narrowing of thought under pressure.

What appears as inconsistency is often not a failure of knowledge, but a system that cannot access what it already knows.

This becomes particularly evident in research on stress. Under sustained pressure, tight deadlines, constant decision-making and unresolved tension with others, the brain shifts in ways that prioritise immediate survival over reflective thinking. The prefrontal cortex, which supports decision-making, impulse control and complex reasoning, becomes less effective, while more reactive systems become more dominant (Arnsten, 2009). In such conditions, individuals lose access to the very capacities they rely on to lead well. It is not that those capacities have disappeared. It is that the system can no longer support them in the same way.

At the same time, the role of the body in shaping experience remains widely underestimated. Physical states are often treated as background conditions, secondary to cognition. Yet physiology plays a central role in how we think and feel. Sleep, nutrition, movement and stress all influence cognitive performance and emotional regulation. Matthew Walker's work on sleep, for instance, shows how even modest sleep deprivation can significantly impair attention, mood and decision-making (Walker, 2017). Anyone who has tried to lead after a poor night's sleep knows this not as theory, but as experience. Clarity becomes harder. Patience shortens. Perspective narrows. Small issues feel larger than they are.

What begins to emerge from this perspective is a gap between what leaders are asked to do and what their systems can support. Leadership development often focuses on behaviours, strategies and competencies while giving far less attention to the conditions that make those behaviours sustainable. As a result, leaders may know what effective leadership looks like and still find themselves unable to embody it consistently. The gap between intention and execution becomes a lived experience.

It is at this point that the idea of Integration becomes essential. In the i4 Neuroleader Model, Integration describes the capacity of the human system to come into coherence, where brain, body and mind are no longer pulling in different directions, but working as one coordinated whole. When that coherence is present, leaders gain access to clarity, steadiness and choice. When it is not, even highly capable individuals can find themselves fragmented under pressure.

Daniel Siegel describes integration as the linking of differentiated parts into a functional system, a process that supports flexibility, resilience and adaptability (Siegel, 2012). When Integration is present, leaders are better able to remain stable under pressure, access a broader range of responses and act in alignment with their values.

When Integration is absent, fragmentation appears. A leader may think one thing, feel another and act in a third way. They may strive for clarity yet experience confusion, seek connection yet withdraw, or aim for composure yet become reactive. They may know exactly what matters and still behave in ways that contradict it.

This fragmentation is often interpreted as inconsistency or lack of discipline. More often, it reflects a system that is not fully integrated. Integration is not a fixed state. It fluctuates with stress, energy, environment and internal patterns. But it can be developed, and it forms the foundation on which effective leadership is built.

None of this is new. What is new is the environment we are trying to function in. The leaders we see today are not failing because they lack knowledge. They are faltering because the world is asking their nervous systems to do something the human nervous system was not built to do.

Leaders are required to process more information, make faster decisions and operate in environments that are increasingly complex and uncertain, often with limited time for reflection or recovery. The conditions that support clarity, stability and sustained attention are becoming harder to maintain.

And now there is artificial intelligence. We are not going to pretend we know exactly where it leads. What we do know, because we are watching it happen in real time, is that it is changing how people think before most of them have noticed. The machines are not replacing cognition. They are absorbing parts of it. These developments offer undeniable advantages, but they also introduce new pressures. As external systems become more powerful, there is a growing risk that the internal system is simply neglected.

The human organism, however, has not evolved for continuous acceleration. It still depends on cycles of activation and recovery, on periods of focused engagement and meaningful rest. When these rhythms are disrupted, the system begins to strain. Attention scatters, emotional regulation weakens and clarity diminishes. In such conditions, leadership becomes more difficult, not because leaders lack ability, but because the system that supports those abilities is under sustained pressure.

This brings us back to the central idea of this chapter. Leadership is not something we construct in the moment. It is something we reveal. Every action, every decision and every interaction reflects the state of the system from which it arises. Before a leader speaks, the system is already communicating, often more powerfully than the words that follow. Before a decision is made, the system has already shaped what is perceived as possible, relevant and necessary.

Leadership must therefore be understood beyond behaviour. Not because behaviour does not matter, but because it does not originate where we often assume it does. Behaviour is the visible expression of processes already in motion, processes unfolding within the human system itself. If we continue to focus only on what is visible, we will continue to overlook what is decisive. We will try to change leadership from the outside in, addressing behaviour without transforming the conditions that produce it.

But leadership does not begin at the surface. It begins beneath it, within the dynamics of a system that must be understood before it can be developed.

Reflection Question

Think of a recent moment when you were under pressure.

Before you said anything, before you made a decision, what was already happening in your body?

And if you had paused for ten seconds to notice that, would the next thing you said have been different?

CHAPTER 2

THE BIOELECTRIC HUMAN

WE ARE ELECTRIC

If the previous chapter invited us to move beyond the idea of the brain as an isolated organ, this chapter asks us to take a further step and reconsider the nature of the human system itself. It is one thing to recognise that the brain does not operate alone. It is another to begin to understand what the system actually is, and what it is made of.

For most of our lives, we have been taught to think about the body as a chemical story. Hormones, neurotransmitters, enzymes. This is how medicine is taught. This is how we were taught. But the body is not only a chemical system. It is also an electrical one, and the electrical dimension has been quietly waiting for its turn. This language has given us extraordinary insight into how life functions and how dysfunction emerges. It has helped us understand how thoughts are influenced, how emotions are regulated and how behaviour is shaped through biochemical processes. Yet chemistry, however essential, does not fully explain the system we are trying to understand. Beneath this chemical layer lies another form of organisation, one that is more immediate, more dynamic and in many ways more fundamental. That layer is electrical.

Every cell in the human body maintains a difference in electrical charge across its membrane. This gradient is sustained through the continuous exchange of ions such as sodium, potassium and calcium through specialised channels and pumps. It is this balance of charge that makes signalling possible. In neurons, these gradients generate the impulses through which information travels. In muscles, they enable contraction.

In the heart, they coordinate rhythm. Electrical activity is not an accessory to biological life. It is one of the conditions that make life possible.

Stop for a moment and picture this. Thirty trillion cells, give or take. Every one of them holding its own electrical charge, every one of them exchanging ions across its membrane thousands of times a second, all of it happening right now while you read this sentence, none of it noticed. That is the hidden magnitude of being alive.

These exchanges occur in milliseconds, outside conscious awareness, yet they underpin everything we perceive, feel and do. Beneath every thought, every emotion and every action lies a constant flow of electrical signalling, a dynamic pattern of charge and exchange that allows the system to function as a whole.

Recognition of this layer has begun to reshape how biology is understood. In *We Are Electric*, Sally Adee brings the emerging science of bioelectricity to a wider audience, drawing together research that had largely remained in specialist journals and showing how electrical gradients function as organising forces within living systems. Rather than being limited to neural communication, electrical signalling appears to play a broader role in how cells coordinate, communicate and maintain structure across tissues (Adee, 2023). This does not replace chemistry. It complements it. Electrical and chemical processes operate together, shaping how the system regulates itself from moment to moment.

To see the human being in this way requires a shift in perspective. We are not static structures moving through the world. We are dynamic systems, continuously organising and reorganising through patterns of signalling that operate beneath awareness. What we experience as stability is, in reality, a continuous process of regulation.

And yet most people live as though none of this were relevant. We push the system beyond its limits, often without noticing. We override fatigue, ignore stress signals and treat the body as secondary to performance. We organise our lives around output, productivity and achievement, assuming that cognition can compensate indefinitely for a system under strain. We behave as though thinking were the primary driver of performance and everything else were optional. But the system does not respond to intention alone. It responds to condition.

For a time, the body adapts. It compensates, recalibrates and absorbs pressure with remarkable capacity. But over time the signals begin to change. Energy becomes less stable. Attention fragments more easily. Emotional responses become less predictable. What is often interpreted as a lack of discipline or resilience may instead reflect a system that is no longer functioning in a coordinated way.

This becomes clearer when we consider how biological systems organise themselves more broadly. Research in developmental biology has shown that cellular behaviour is influenced not only by genetic instructions, but also by patterns of signalling that guide how cells grow, differentiate and repair. Bioelectric states appear to contribute to this process, helping maintain structure and function across living tissues (Levin, 2012). What we observe at the surface is often the result of deeper patterns already in motion.

The same principle applies to human experience. Thoughts do not arise independently. Emotions do not appear in isolation. Behaviour is not generated in a vacuum. All of these emerge from processes through which the body and brain continuously interpret and regulate internal and external signals. Research on interoception provides an important bridge here. The brain is not only responding to the external world.

It is constantly modelling the internal state of the body and using that information to construct perception and meaning. What we experience as emotion, intuition or even judgement is shaped through this ongoing integration of internal and external signals (Barrett & Simmons, 2015).

From this perspective, experience is not simply reactive. It is interpretative, and that interpretation is grounded in the condition of the system itself. This has direct implications for how we understand clarity, confusion, calm or reactivity. These are not only psychological experiences. They are expressions of how the system is organised in a given moment. When internal signalling is more coherent, experience tends to be more stable, flexible and responsive. When signalling becomes fragmented, experience narrows and the range of available responses contracts. Before performance changes, possibility does.

Most people know this without using the language for it. The day when patience shortens for no obvious reason. The meeting that feels heavier than it should. The moment when thought becomes narrow, not because the problem changed, but because the system did.

This is also central to leadership. When a leader enters a room, they bring more than ideas, expertise or authority. They bring their system. Their state influences the environment, shaping how others feel, respond and engage. This influence is not only cognitive or verbal. It is physiological. Human systems are highly responsive to one another, and they tend to align through subtle processes that occur largely outside conscious awareness. In this sense, leadership is not only internal regulation. It is also transmission. The condition of one system influences the condition of another.

To develop leadership, therefore, is not only to refine behaviour. It is to understand and work with the system that produces it.

What makes this practical is not the introduction of new tools, but the development of a different kind of awareness. Most leaders are trained to focus on outcomes, targets and performance indicators. Few are trained to notice the internal signals that precede those outcomes.

And yet the earliest indicators of performance are not external. They appear within the system itself. They show up in the moment attention begins to drift, even when the task remains important. They appear when patience shortens in conversations that would normally feel manageable. They appear when urgency replaces clarity and the ability to see broadly begins to narrow without being consciously recognised. These are not failures. They are signals.

The difficulty is that these signals are often overridden. Modern environments reward continuity, speed and output. Leaders learn to continue despite internal feedback, to push through strain and to maintain performance even when the system is beginning to lose coherence. Over time this creates a disconnect between what is experienced and what is acknowledged. What is no longer noticed begins to feel normal. Interpretation becomes distorted. Fatigue is seen as lack of discipline. Reactivity is framed as a mindset issue. Inconsistency is attributed to insufficient effort. In many cases these experiences are not problems of intention. What looks like weakness is often strain that has gone unnamed.

To work with the system rather than against it requires the ability to notice earlier in the sequence. Not after the reaction, but before it. Not when the system is already overwhelmed, but when it begins to shift. This does not require significant time or complex techniques. It requires brief moments of awareness within the flow of activity. A pause before responding, not to analyse but to sense. A moment of attention to breathing, not to control it but to notice its rhythm. A recognition of tension in the body, not as something to eliminate, but as information about the state of the system.

These moments are small, but they matter. When awareness enters earlier in the process, the system gains options. Automatic reactions are no longer the only pathway available. Regulation can occur before escalation rather than after it. Over time, this changes how leadership is experienced, not as an effort to maintain control, but as a capacity to operate from a more coherent condition. Decisions begin to reflect not only strategy, but state. Conversations are shaped not only by content, but by presence. Performance becomes not only a function of output, but of the sustainability of the system that produces it.

This is not about slowing leadership down. It is about reducing unnecessary friction within the system that is trying to lead. When the system becomes more coherent, less energy is spent compensating for internal misalignment. Attention stabilises, emotional range expands and thinking becomes both clearer and less effortful. What once required control, discipline or force begins to emerge with greater ease, not because the demands have disappeared, but because the system is no longer working against itself.

From the outside, this is often described as composure, clarity or presence. These are the visible expressions. From the inside, however, the experience is different. It is not a performance. It is a system functioning in a more coordinated way, allowing perception, emotion and action to align rather than compete. The leader is not trying harder. The leader is working from a condition that supports what needs to happen.

If leadership begins beneath behaviour, then it begins within a system whose condition determines what becomes possible in any given moment. The question, then, is no longer only what you do as a leader. It is what your system allows you to do. Moment by moment. Under pressure. In complexity. In the presence of others. That is where leadership is formed. And that is where it can be developed with far greater precision than behaviour alone will ever allow.

Reflection Question

If you paid closer attention to your internal state for just one day, not to change it, but simply to notice it, what patterns would begin to reveal themselves that you normally move past without seeing?

And once you noticed those patterns, what would change about how you walked into your next conversation?

CHAPTER 3

LEADERSHIP AS A LIVING CIRCUIT

LEADERSHIP IS TRANSMISSION

If the previous chapter invited us to understand the human being as bioelectric, this chapter asks what that means in lived experience. It is one thing to recognise that we are electrochemical organisms whose cells are continuously exchanging signals. It is another to realise that this signalling does not remain sealed inside us as a private biological event. It shapes the way we perceive, feel, regulate and influence one another. Once we begin to see the human system as an integrated network rather than a hierarchy with the brain at the top, leadership itself begins to look different.

It is no longer sufficient to describe leadership as decision-making, strategic thinking, persuasion or behavioural competence. Those things still matter, but they are not the deepest layer of what is happening. Leadership is not only the communication of ideas. It is the transmission of state.

For a long time, leadership theory has focused on visible action. Leaders are expected to articulate vision, make sound judgements, inspire confidence, regulate emotion and guide others through complexity. Yet anyone who has spent enough time around teams, organisations or families knows that influence begins long before words do. We feel people before we fully understand them. We sense whether someone is steady or fragmented, defensive or open, grounded or performative. A leader does not begin to influence others when speaking starts. Influence begins the moment another nervous system encounters their presence. This is not mystical. It is biological.

Human beings are not only thinking beings. We are sensing, signalling and synchronising beings. The brain participates in leadership, but not alone. The heart participates. The gut participates.

The autonomic nervous system integrates all of it, while the body continuously expresses the condition of the system to the world around it. Leadership, from this perspective, is not simply what a person thinks and does. It is what the whole circuit is transmitting.

The heart offers one of the clearest examples of why this matters. In conventional medical language, the heart is often described as a pump, which is true as far as it goes, but incomplete. Research on heart coherence has shown that cardiac rhythms are closely linked with emotional regulation, attentional stability and self-regulation (McCraty & Zayas, 2014). Bruce Cryer, former CEO of HeartMath LLC, has pointed out something uncomfortable here. In business, we often dismiss the heart as sentimental, while in elite sport we openly recognise that trust, cohesion and emotional steadiness determine performance. The biology is the same. Only the narrative changes. What becomes clear is that the heart is not simply following instructions. It participates in regulation. It helps shape how the body and brain coordinate in real time. The condition of the heart is not only a medical issue. It is part of the architecture of presence.

You can see this play out in everyday leadership more clearly than in any theory. A senior executive walks into a meeting to approve a major strategic move. On paper, everything aligns. The numbers are strong. The logic is sound. The case has been prepared thoroughly. And yet somewhere in the body something feels off. A subtle contraction. A hesitation that cannot be explained in a spreadsheet. It is noticed briefly, then dismissed. "We do not have time for doubt." The decision goes ahead. Months later, the issues that were felt but not named begin to surface. Misalignment. Resistance. Unintended consequences. And the sentence appears, almost always too late: "I had a feeling." This is not a failure of intelligence. It is a failure of listening.

In another moment, a leader sits across from a colleague who is struggling. The conversation is factual, efficient, professional. Targets are discussed. Solutions are proposed. The leader senses something deeper, a moment where slowing down, asking a different question or simply staying present might open something important. But the moment passes. The agenda takes over. The meeting ends. Nothing that was said was wrong, and yet something important was missed.

These are not rare situations. You have lived them. You have felt the moment when your body registered something before your mind caught up. You have overridden it because it did not fit the pace, the culture or the expectation of being rational. You have stayed in analysis when something in you was asking for attention, connection or pause. This is where leadership becomes deeply human and often silently compromised.

Dr Jonathan Fisher, cardiologist and author of Just One Heart, has observed this in high-pressure environments where performance appears intact while internal strain continues to accumulate. People follow the right behaviours. They meet expectations. They comply with what is required. And yet the emotional and relational dimensions of their lives remain unaddressed. Over time, the system begins to show the cost, not because people do not know what to do, but because they are no longer connected to the signals that would guide them differently. The cost is rarely immediate. It accumulates.

This has profound implications for leadership. A leader may look composed and capable while internally carrying tension that is constantly transmitted. They may speak about trust while their system communicates urgency. They may promote collaboration while their presence signals pressure or distance. This is one reason teams respond less to what leaders say than to what leaders are. Calm, in this context, is not a personality trait. It is a regulatory state. It shapes how others think, feel and respond.

The steadiness of a leader does not remain contained within them. It becomes part of the environment. The same is true for anxiety, impatience or defensiveness. These states travel.

Judith Glaser's work on conversational intelligence made this visible in practical terms. Conversations are not neutral exchanges of information. They are biological events that influence whether people move into trust or protection. When communication triggers fear or judgement, thinking narrows and collaboration declines. When it creates safety and openness, a different level of thinking becomes available (Glaser, 2014). What looks like a communication issue is often a state issue.

The challenge is that modern work cultures reward the opposite of what the human system needs. Speed is rewarded over presence. Control over awareness. Certainty over sensing. Leaders learn, often unconsciously, to live in their heads. They override signals from the body because those signals are not measurable, immediate or convenient. Over time something subtle begins to happen. The body continues to communicate, but the leader stops listening.

Dr Kaushik Ram has described this as a form of instinctual illiteracy, where people become highly capable thinkers while losing fluency in their own internal signals (Ram, 2024). It is not that intuition disappears. It is that it is no longer recognised as valid input. And yet the body does not stop participating. The heart continues to respond in real time. It reflects shifts in attention, emotion and meaning through changes in rhythm and physiological tone. It does not wait for analysis. It registers what is happening now.

Ram's work points to something simple but often overlooked. When people begin to notice these signals earlier, through breath, tension or subtle changes in activation, they can interrupt the cascade that turns physiological activation into reactive thought.

This interruption matters because many leadership failures are not failures of capability. They are failures of timing. The moment when the system could have shifted passes unnoticed. The signal is overridden. The reaction becomes inevitable.

The same applies to what we casually call a gut feeling. Dr Emeran Myer's work on the gut-brain axis has shown that the digestive system is deeply involved in emotional and cognitive processes, communicating continuously with the brain through neural and biochemical pathways (Mayer, 2016). What people describe as intuition is not abstract. It often reflects the body's capacity to detect patterns before the conscious mind has articulated them. And yet, in many organisations, this kind of knowing is dismissed. A leader senses that a partnership feels wrong but proceeds because the data supports it. Another hesitates before hiring someone but overrides the hesitation because the candidate looks perfect on paper. In both cases, the decision is rational. In both cases, something was known earlier and ignored. This is not about replacing analysis with instinct. It is about recognising that analysis without sensing is incomplete.

One useful way of understanding this is to see that human intelligence is not expressed through thought alone. The head contributes analysis and perspective, the heart connects us to values and relationships, and the gut relates to action, boundaries and commitment. When these dimensions are aligned, behaviour becomes more coherent and grounded. When they are not, leaders tend to overthink, disconnect or act in ways that later feel misaligned (Soosalu & Oka, 2012). This alignment is not theoretical. It is experienced. You can feel when you are aligned. Decisions feel cleaner. Actions feel grounded. There is less internal negotiation. And you can feel when you are not. Something drags. Something resists. Something feels forced. Most leaders have learned to push through that signal.

The brain, in all of this, is not the commander of passive systems. It is part of a continuous exchange. It interprets, predicts and integrates, but it does so in constant dialogue with the rest of the organism. Models of neurovisceral integration highlight how closely cardiac regulation, attention and adaptive response are linked (Thayer & Lane, 2009). What we call thinking is already shaped by processes happening throughout the body. By the time a leader speaks, the system has already been active.

This is why leadership is transmission. It is the transmission of a pattern, not just a message. A leader walking into a room brings a way of being that others respond to immediately. If that pattern is coherent, it creates stability and trust. If it is fragmented, it creates tension and uncertainty. This happens before any strategy is explained or any direction is given. It is also why so many leaders feel a form of disconnection over time. They become highly capable at performing leadership while feeling less connected to themselves in the process. They rely more on analysis and less on sensing. They make decisions faster, but often with less internal alignment. They succeed externally, while something internally becomes less clear.

This chapter is not suggesting that leaders abandon rational thinking. It is suggesting that rational thinking was never meant to operate alone. Some of the most important dimensions of leadership occur beneath language. They occur in the body, in the rhythm of breathing, in the variability of the heart, in the signals of the gut and in the way these shape what is felt in relationship. To ignore this is not only to miss part of the science. It is to miss part of what makes leadership human.

If modern life continues to train people away from instinct and into continuous mental activity, then leadership becomes one of the places where that pattern must be interrupted. Not by rejecting technology, but by refusing to let it replace the body as the primary reference point for being human.

Not by dismissing cognition, but by restoring it to its place within a wider system. Because the future of leadership will depend not only on what we know, but on whether we can remain connected to the signals that allow us to use that knowledge wisely. That begins with recognising the circuit we already are.

Reflection Question

Think of a recent moment where you made a decision quickly.

Was there anything you felt before you decided
that you chose to ignore?

What might have changed if you had stayed with that
signal just a little longer?

CHAPTER 4

THE STATE BEFORE THE STRATEGY

A SHIFT IN PERSPECTIVE

This may sound like a subtle shift in emphasis, but in practice it changes everything. For decades, leadership has been developed largely from the outside in. We have taught people what effective leaders do, how they speak, how they make decisions, how they influence others and how they respond to challenge. There is value in all of this. Good frameworks can clarify what matters. Good models can sharpen attention. Good practices can help people move beyond instinctive habits that no longer serve them. But beneath all of that sits an assumption that has gone largely unquestioned: if people understand what effective leadership looks like, they will be able to reproduce it consistently when it matters most.

Experience suggests otherwise. A leader may know how to communicate with clarity and still become abrupt under pressure. They may understand the importance of listening and still interrupt. They may value trust, collaboration and emotional steadiness, yet find themselves narrowing, hardening or withdrawing precisely when those qualities are needed most. It is not that the knowledge disappears. Access to it becomes compromised. In those moments, what leads is not aspiration, but state. Not intention, but the condition of the system.

This is where the conversation becomes more personal than many leadership books allow. Because if we are honest, most of us have lived this not only at work, but at home. We have all had moments in which we wanted to be generous and were instead impatient, wanted to be calm and were instead reactive, wanted to be present and were already somewhere else internally. That is part of what makes this inquiry so important.

It applies not only to formal leadership roles, but to how we show up in relationships, in families, in conflict, in care and in the quiet exchanges that shape a life.

Behaviour alone is never the full story. It is the visible surface of a deeper process, the final expression of something already unfolding within the body and nervous system before the leader has spoken a word. A strategy may be well designed, a message carefully crafted, and a set of leadership values sincerely held, yet all of it will be shaped in the moment by the state from which it is delivered. This is what makes leadership both more demanding and more human than many traditional models have allowed.

Part of the difficulty is that we have inherited a view of effectiveness that overvalues cognition and undervalues condition.

We assume that if we can think clearly enough, prepare thoroughly enough and discipline ourselves sufficiently, we will be able to lead well. But human beings are not organised that way. The mind does not float above the rest of the organism issuing commands to an obedient body. It emerges within a living system that is continuously regulating, adapting, predicting and responding. The quality of that system matters, and it shapes far more than many leadership theories have been willing to admit.

Across years of working with leaders, we have observed something that sits just beneath the surface of every leadership conversation. Sustained effectiveness does not depend only on cognitive ability or managerial skill. It depends on whether a leader can access the internal states that allow openness, connection and adaptive functioning to remain available when they are needed. We have watched highly intelligent leaders lose access to their own capabilities under pressure, not because they forgot what to do, but because the system from which they were leading could no longer hold it.

We have also watched other leaders, sometimes less credentialed, move through the same conditions with clarity, empathy and flexibility intact.

The difference was never about knowledge. It was about state. Leaders do not perform at their best simply because they know more. They perform at their best when the system from which they are leading allows clarity, empathy, flexibility and judgement to remain available under real conditions. What separates these two modes is rarely visible from the outside, yet it shapes everything that follows. One produces leadership as performance. The other produces leadership as embodied capacity. One generates temporary compliance. The other generates sustainable transformation.

A person can imitate a behaviour for a while. They can rehearse it, deliver it and even be praised for it. But if the system underneath remains fragmented, strained or chronically dysregulated, the behaviour will not hold. Under pressure, the system reverts to what it knows. This is why so many leaders find themselves repeating the same frustrations. They attend development programmes, understand the principles, commit to change and still return to familiar patterns when stress rises. The issue is rarely a lack of commitment. More often, the internal architecture required to sustain the new behaviour has not yet been established. Under pressure, performance gives way to pattern.

To work with leadership more truthfully, we must become more interested in what happens beforehand. We must become more attentive to what happens in the seconds before we respond, what happens in the body when pressure rises and what happens in the nervous system when we feel challenged, observed, rushed or uncertain. These moments often appear small, but they are not. They are the hinge on which behaviour turns. The leader who learns to recognise them begins to lead from a different place altogether.

This does not mean becoming hyper-analytical or self-absorbed. In fact, the opposite is true. It means becoming simpler and more precise in one's awareness. Ellen Langer's work on mindfulness is useful here, not because this chapter is about meditation in any narrow sense, but because she describes mindfulness as the active process of noticing what is actually happening rather than operating automatically from old categories and assumptions (Langer, 1989). In leadership, this kind of noticing matters immensely. It allows a person to recognise when they are no longer responding to the present moment, but to a pattern. It creates a pause, however brief, in which a different response can emerge.

Herbert Benson's work on the relaxation response mattered because it showed, in language people could understand, that the human body is not condemned to remain in cycles of stress once activated. It can move toward states that counter chronic physiological arousal and restore a greater sense of balance (Benson, 1975). That matters here because leadership is not simply about managing behaviour after the fact. It is about recognising the conditions from which behaviour arises and learning to work with them earlier and more intelligently.

There is another reason this matters, and it gives this inquiry both weight and tenderness. Leadership does not happen alone. It unfolds between human beings, and human beings are profoundly shaped by one another's states. Emotional contagion research has long shown that people influence each other's emotional and physiological condition through interaction, often without deliberate intent (Hatfield, Cacioppo, & Rapson, 1994). In practical terms, regulation is never merely personal. It is relational. A leader who enters with chronic urgency does not simply experience that urgency privately. They distribute it. A leader who enters with steadiness and warmth does the same. These states become part of the wider field.

This is why presence matters so much, and why it has often been misunderstood. Presence is not charisma. It is not theatrical confidence or verbal fluency. Presence is the degree to which a leader's system is coherent enough that others can settle around it rather than brace against it. Presence widens thinking in the room. It allows trust to grow more quickly. It makes disagreement more workable and complexity more thinkable. In professional life this can change the quality of a team, the tone of a difficult meeting or the outcome of a strategic decision. In personal life it can change the atmosphere of a home, the safety of a conversation and the emotional inheritance we pass to those closest to us. Presence is not performance. It is regulation made visible.

This is where the idea of Integration becomes essential, and where the i4 Neuroleader Model brings conceptual clarity to what Part I has been pointing toward. Within the model, Integration is one of the foundational human abilities. It refers to the capacity of the system to function as a coherent whole rather than as a collection of competing parts. It is not a technique to perform, but a condition to develop. Without Integration, even the most refined leadership intentions struggle to take form, because the system lacks the internal coordination required to sustain them.

As Part I draws to a close, the invitation is simple. Begin to notice the state from which you are leading, not obsessively, not analytically, but honestly. Notice when you are bracing, when you are rushing, when you are contracting, when you are settled. Notice what your system is offering you in the moments before you act. That noticing is where the next phase of this book begins, because once we understand the system beneath leadership, we are ready to ask what happens when that system is placed inside a world moving faster than it was built for.

Reflection Question

Think about the next important conversation you are about to have.

Before it begins, ask yourself one simple question:

What am I bringing into this moment that others are likely to feel before I even speak?

PART II

A WORLD MOVING FASTER THAN THE HUMAN SYSTEM

WHAT WE ARE BUILDING EXTERNALLY IS RESHAPING US INTERNALLY, WHETHER WE ARE AWARE OF IT OR NOT

CHAPTER 5

THE DYSREGULATED LEADER

WHEN THE SYSTEM CANNOT HOLD

Leadership rarely breaks in dramatic moments. It erodes in ordinary ones. A morning that begins with pressure. A reaction that comes faster than intended. A tone that lands harder than expected. Nothing appears significant in isolation, yet something in the system has already shifted. Before the day has fully begun, the body is already carrying more than it can regulate. This is where Chapter 5 begins, not in collapse, but in the small shifts that precede it.

A father stands in the kitchen, already slightly behind the day. His phone has been active since early morning. Messages, emails and reminders have begun to accumulate before he has even left the house. His mind is already moving ahead to what needs to be done. A meeting at nine. A client waiting. Work that feels urgent before it has even begun. Nothing dramatic has happened, yet his system is already activated. The day has not started, but the pressure has.

From another room, his wife asks whether he can drop the children at school. The request is simple. It is not unreasonable. But something in him shifts almost immediately. Not in thought, but in the body. His breath shortens. His chest tightens. Attention narrows toward time, pressure and interruption. He agrees, but the system has already reorganised around urgency.

The children are not ready at the pace he needs them to be. One is looking for something. The other is distracted. Shoes are not on. Bags are half packed. These are small, predictable moments in any household, but in the state he is in, they are experienced differently. They are no longer neutral.

They are obstacles. His voice rises. He tells them to hurry up. The urgency he feels becomes the urgency they are expected to carry. They respond quickly. They say yes. They move faster. But something changes in them as well. Their bodies register the tone before they fully understand the words. The day begins not with steadiness, but with pressure. No one names it. The moment passes. The state does not.

He drops them at school and drives away already aware, somewhere beneath the surface, that the interaction did not reflect the kind of parent he believes himself to be. That awareness is brief. It competes with the demands of the day. By the time he arrives at the office, he has moved on cognitively, but not physiologically.

At nine o'clock, he is in a meeting. He is present, but not fully available. His attention is narrower than usual. His patience is reduced. He listens, but not deeply. He responds, but with a slight edge. His tone carries something he does not intend, yet others feel it immediately. No one may challenge it directly, but the atmosphere shifts. Contributions become more cautious. The conversation becomes more transactional. Something in the room tightens.

Later, in a coaching session, the same pattern appears again. Irritation surfaces faster than expected. He interrupts. He pushes harder than necessary. He frames it internally as efficiency, as maintaining standards, as doing what needs to be done. And yet, beneath that justification, there is a recognition that something is not aligned. He knows this is not how he wants to lead. He knows it does not bring out the best in others. But in the moment, he cannot access a different response. This is not a story about character. It is a story about condition.

What unfolds across that morning reflects a system operating under sustained pressure. When stress levels rise, the brain does not simply continue functioning at the same level with added intensity. It reorganises. Neural resources shift away from the prefrontal cortex, the region responsible for reasoning, impulse control and flexible thinking, toward more reactive circuits designed for speed and survival (Arnsten, 2009). The leader does not lose intelligence. They lose access to it in the moments when it is most needed.

This shift is not a failure of discipline. It is a biological adaptation. Under conditions of perceived pressure, the system prioritises efficiency over reflection, speed over nuance and action over consideration. In short bursts, this can be useful. It allows rapid response. It supports decisive movement. But when this state becomes prolonged, it begins to narrow what the leader is able to perceive, consider and hold.

Over time, the effect becomes cumulative. The body absorbs the load of continuous activation, and what was once a temporary state becomes more stable. Energy becomes uneven. Recovery becomes slower. Emotional responses become faster and more difficult to regulate. This process has been described as allostatic load, the cumulative cost of chronic stress on the organism (McEwen, 1998). It is not a sudden collapse. It is a gradual shift in baseline, where strain becomes normal and regulation becomes harder to access.

Many of the leaders we work with describe themselves the same way. Always on. Always thinking. Alert but never quite at rest. They tell us they are functioning, and they are. But when we ask them what it costs, they stop and think for a moment, and then say something that surprises even them. It costs everything else. Their weekends. Their patience. The quality of their presence with their children. The texture of their inner lives.

We have heard all of these explanations, usually from the leaders themselves. "I am just direct." "I am under a lot of pressure." "This is what leadership looks like at my level." None of these are lies. They are the stories the nervous system tells itself to justify what it can no longer prevent. What is less visible is that the system has moved outside its optimal range.

When the human organism perceives pressure or threat, it shifts toward protection. Attention narrows. Social engagement reduces. The capacity for connection becomes secondary to the need for control or efficiency. This shift is not deliberate. It is regulated through the autonomic nervous system, which continuously scans for cues of safety or threat and adjusts the system accordingly (Porges, 2011). The leader is not choosing disconnection. The system is prioritising survival over relational presence.

At the same time, the emotional dimension of leadership becomes compromised. When emotional signals are not processed or integrated, they do not disappear. They accumulate. Over time, that accumulation shapes how individuals interpret situations, how they perceive others and how they respond under pressure. Research on trauma and stress has shown that unresolved patterns of activation can remain stored in the body, influencing behaviour outside conscious awareness (van der Kolk, 2014; Maté, 2019). What appears as overreaction in the moment often reflects a system that has been carrying more than it has processed. The moment is only the surface. The load is older.

This is where dysregulation extends beyond the individual and begins to shape the environment. When people describe a good leader, they rarely speak in terms of frameworks or competencies. They describe how it feels to work with them. They speak about clarity, trust and being able to do their work without unnecessary pressure. They describe someone who understands the work, makes decisions when needed and allows others to take ownership without constant interference.

In one conversation, an employee described her manager in exactly these terms. He was knowledgeable and able to process information quickly, yet he did not micromanage. He trusted her to do her work and checked in when needed. His feedback was simple and timely rather than heavy or disruptive. When personal challenges arose, he responded with consideration rather than resistance. There was no sense of being managed through pressure. There was a sense of being supported to function well. What stands out in this description is not what he adds, but what he does not create. He does not introduce tension into the system. He does not amplify stress. He does not make people smaller in order to move faster. This is not a soft quality. It is a structural one.

Human beings are highly sensitive to one another's internal state. A leader who operates from chronic pressure does not contain that pressure. It becomes part of the environment. It shapes how others think, how they prioritise and how safe they feel to contribute. Over time, this has measurable effects on performance and learning. When people do not feel psychologically safe, they are less likely to speak up, share ideas or admit mistakes, which directly limits collective intelligence (Edmondson, 1999). The opposite is also true.

When the leader's system is more regulated, others gain access to their own cognitive and emotional capacity more easily. Emotional regulation is not only an individual capability. It becomes a shared condition that influences how groups function, collaborate and perform. Research in emotional intelligence has shown that environments where emotions are understood and regulated effectively are associated with stronger performance, better decision-making and healthier working relationships (Brackett, 2019).

When regulation is absent, leaders often compensate through control. We have seen this pattern many times. Underlying anxiety, particularly anxiety related to performance or status, drives behaviours that appear decisive on the surface but are rooted in internal instability. A leader who has lost access to their own steadiness begins to manage that loss outwardly. Increased monitoring. Reduced tolerance for difference. Sharper communication. Faster decisions. Less room for challenge. From the outside, these look like signs of a leader who is in command. From the inside, they are often signs of a leader who is unravelling and trying to hold the ground beneath them through force rather than presence.

We have watched this in senior executives whose teams described them as intimidating or volatile without ever connecting those behaviours to the person's own internal strain. We have watched it in founders who built their companies through vision and conviction and who began, years later, to micromanage the same people they had once trusted. We have watched it in middle managers who were given responsibility without the conditions or support to hold it, and who responded to the gap by becoming harder rather than more open. In every case, the behaviour was not character. It was compensation. The leader was not trying to be difficult. They were trying to manage an internal pressure they could no longer regulate any other way.

These behaviours may produce short-term results. They may create movement. They may even be rewarded in the early stages, because they can look like focus, drive or accountability. But over time, they erode trust and engagement in ways that become visible too late to reverse easily. People adapt to them, but they do not thrive within them. The team becomes voiceless. Ideas stop surfacing. Mistakes get hidden rather than corrected. And the leader, who began by trying to hold everything together, ends up holding less and less, because the system around them has learned to protect itself from the very person it was meant to follow. Control may keep the machine moving for a while. It does not keep the human system alive.

This is how dysregulation becomes embedded in leadership culture. It is not always loud or visible. It can exist in subtle patterns of tension, urgency and reactivity that shape how work is experienced every day. It lives in tone, timing, the absence of patience and the presence of pressure. It becomes normalised, not because it is effective, but because it is repeated.

At a more fundamental level, what people are responding to is not behaviour alone, but the state beneath it. They are responding to whether they can think without pressure distorting their attention, whether they can contribute without navigating unnecessary tension and whether they can engage without anticipating reaction. These are not abstract preferences. They are conditions that allow the human system to operate effectively. When those conditions are present, individuals expand. Attention stabilises. Thinking becomes more flexible. Collaboration becomes more natural. When they are absent, the opposite occurs. Attention narrows. Responses become more reactive. Contribution becomes more cautious. Learning slows. Innovation diminishes. This shift often happens before anything is explicitly said or done. It happens at the level of state. Before culture becomes visible, it is felt.

Dysregulation matters so profoundly in leadership not because it makes leaders imperfect, but because it shapes what others are able to access in themselves. A leader whose system is under continuous strain may still perform, still deliver and still move forward, but the cost is carried across every interaction. If the system cannot hold, the leader cannot consistently hold others. And if that layer remains unexamined, no amount of strategy, skill or intention will fully compensate for what the system is unable to sustain.

Reflection Question

Think about a recent moment when you reacted more sharply than you intended.

What was happening in your body just before that reaction, and did you notice it at the time?

CHAPTER 6

FRAGMENTED ATTENTION, FRAGMENTED SELF

WHEN THE MIND IS EVERYWHERE AND NOWHERE

If pressure disrupts the system, fragmentation prevents it from recovering. What begins as urgency in the body becomes instability in attention. The mind no longer settles. It moves, reacts, switches and scans, but rarely arrives. We are not losing attention. We are losing the ability to remain with it. And without that, something deeper begins to fragment.

The scene is no longer unusual, which is precisely why it deserves attention. A family sits at a breakfast table on a Sunday morning, a moment that once might have offered pause before the rhythm of the week resumed. Instead, the father's attention is fixed on the iPad in front of him. The children are each absorbed in their own devices, enclosed within separate streams of stimulation. A waiter approaches to offer assistance, speaks once, then again, and still no one looks up. Only when the voice becomes slightly louder does one of the children lift their eyes and say, "Dad, they're talking to you."

What makes the moment troubling is not its drama, but its familiarity. Nothing visibly catastrophic has happened, and yet something is absent. Presence has been replaced by partial attention. A shared human moment has been overtaken by competing demands on the mind, each one small enough to seem harmless, yet together strong enough to fracture the whole.

The same pattern appears, with different furniture and different stakes, in workplaces everywhere. A meeting begins and everyone is physically present, yet attention is scattered across screens, internal dialogue, incoming messages and anticipatory thought.

One person is speaking while another is half listening and half drafting a response to an email. Someone else is preparing their next point instead of hearing the current one. Another glances at a notification and returns to the room, but only partially. The meeting continues, decisions are made, and the surface appearance of productivity remains intact. Yet the quality of attention in the room has changed, and with it the quality of thinking. This is where the real problem begins to reveal itself. Attention has not disappeared. It no longer settles. It has become fragmented.

Research into attention in digital environments shows that the mind does not shift cleanly from one task to another. Each interruption carries a cost that is rarely visible in the moment. Attention does not simply move on. It lingers. Part of the mind remains attached to what came before, creating what researchers describe as attention residue. Over time, this repeated switching consumes cognitive energy and reduces the capacity to remain fully engaged with what is in front of us. What feels like efficiency is often a hidden form of depletion. Gloria Mark's work makes this visible by showing how frequently attention is interrupted and how difficult it becomes for the mind to return to a state of sustained focus once that continuity has been broken (Mark, 2023).

Recent research also suggests that the duration of sustained attention on a single task has been steadily decreasing. While the exact figures vary, the direction is consistent. Attention is becoming shorter, more reactive and less stable over time. This is not because people are less capable, but because the environments in which they operate are training the mind toward constant switching rather than sustained engagement. Over time, the brain adapts to this rhythm. It becomes better at responding quickly, but less comfortable remaining with one thing long enough for depth to emerge.

This matters because leadership requires more than speed. It requires duration. It requires the ability to stay with complexity without immediately resolving it, to listen beyond the first layer of what is being said, and to allow thought to unfold beyond its initial response. These capacities are not abstract. They depend on attention stable enough to remain. When attention fragments, the leader may still appear functional, still communicate and still decide, but the coherence behind those actions begins to thin. Decisions may become quicker without becoming wiser. Responses may become more polished without becoming more thoughtful. Meetings may move faster while producing less insight.

Some of the most meaningful human experiences depend on the opposite of fragmentation. They depend on sustained attention. The state often described as flow, where a person becomes fully immersed in what they are doing, requires continuity rather than interruption. It is in these states that thinking deepens, creativity expands and a sense of meaning becomes available. When attention is repeatedly broken, access to these states diminishes. The leader may remain busy, engaged and productive, yet find it increasingly difficult to experience the depth of focus that allows for real insight or genuine satisfaction.

The problem is not only cognitive. It is also relational. In Alone Together, Sherry Turkle explores what happens when devices are constantly present in human interactions, even when they are not actively being used. Their presence alters the emotional field of conversation. People become less willing to go deeper, less comfortable with silence and less able to remain in the kinds of pauses where reflection, vulnerability and genuine contact often emerge. We are together, but only partially. We are connected, but more thinly. The interaction continues, but its quality changes (Turkle, 2011). For leadership, this matters enormously. A leader who cannot offer undivided attention cannot create the conditions in which others feel fully received.

And when people do not feel fully received, they rarely bring their most careful thinking. They offer what is manageable, what fits the pace of the room, and what can be expressed quickly without risk.

Over time, the fragmentation of attention begins to affect more than behaviour. It affects the experience of self. Attention does not only determine what we focus on. It shapes how we experience continuity within ourselves. A coherent sense of self depends, in part, on the ability to remain with experience long enough to integrate it. When attention is repeatedly divided, that continuity weakens. The self becomes more distributed, less anchored, moving across multiple streams of input without fully settling in any of them. The individual remains active, but less gathered. Part of the mind is always elsewhere, and part of the self is always in transit.

This has implications that go beyond fatigue or distraction. When attention is repeatedly divided, the ability to form a continuous narrative of experience begins to weaken. We rely on attention to connect moments, make sense of what is happening and integrate it into memory. When that process is interrupted again and again, experience becomes more episodic and less coherent. A conversation is remembered in fragments. A decision is made without fully recalling the conditions that shaped it. A day passes, full of activity, yet is difficult to reconstruct in any meaningful way. Over time, this weakens not only memory, but identity itself. The sense of being a consistent, grounded presence in one's own life begins to erode, replaced by a more reactive and dispersed experience of self. Attention fragments. The self follows.

The wider context makes this easier to understand. The environments we now inhabit are not neutral. They are designed to capture attention, hold it and redirect it. They do not require us to disengage completely. We have watched this happen in our own lives. Not as a theory. As a lived reality. The phone that used to be a tool has become something closer to a limb. The hand reaches for it before the mind decides. That is not a character flaw. That is what these systems are designed to produce, and they produce it in us too. This reframes the problem. It is not simply that individuals lack discipline. It is that the systems surrounding them are structured in ways that make sustained attention increasingly difficult to maintain.

The cognitive consequences of this environment become even clearer when we consider how patterns of information consumption shape thought itself. In The Shallows, Nicholas Carr explores how environments built around speed, scanning and constant movement may gradually retrain how the mind works. When information is repeatedly encountered in fragments, thinking begins to mirror that structure. The mind becomes quicker at recognising and reacting, but less comfortable with immersion, sustained reflection and deep synthesis (Carr, 2010). This does not eliminate intelligence. It redistributes it. Certain capacities strengthen, while others become harder to access.

In leadership, this becomes visible in subtle but important ways. A leader walks into a meeting after a morning of continuous interruption. Several emails have already triggered small surges of urgency. Messages have left residues of irritation or concern. Notifications have kept part of the mind partially open to what might come next. The meeting begins. A team member raises a nuanced issue that requires careful attention. It is heard, but not fully processed. The response is quick, perhaps even articulate, but not deeply considered. Something important is skimmed over, not because the leader lacks intelligence, but because attention no longer has the duration required to hold the issue long enough for better thinking to emerge.

A similar pattern appears in one-to-one conversations. A leader sits with a team member who is trying to explain a concern that has been building for weeks. The words are clear, but the leader's attention is divided.

Part of the mind is tracking the conversation. Another part is already thinking about the next meeting. Another is holding on to something unresolved from earlier in the day. The leader listens, nods and responds. From the outside, the interaction appears complete. But the person speaking leaves with a subtle sense that they were not fully heard. Nothing explicit has gone wrong, yet something essential has not occurred. The opportunity for deeper understanding, for trust to strengthen or for the issue to be seen in its full context has been missed, not because of lack of care, but because of lack of presence.

Artificial intelligence introduces another layer to this landscape, and caution is required here. The long-term cognitive consequences are still emerging. However, early evidence suggests that as systems become increasingly capable of summarising, drafting and organising, the temptation to offload more cognitive work will increase. This may improve efficiency while reducing the need to sustain attention through difficulty. When thinking is consistently supported or replaced by external systems, the internal processes that sustain it may be exercised less frequently. The effort required to hold complexity, sit with uncertainty or work through ambiguity may begin to feel unnecessary or even inefficient.

Over time, this can subtly shift expectations. Thinking becomes something to be assisted rather than endured. The discomfort that often accompanies deeper reasoning, the pauses, the friction, the slow formation of insight may be avoided rather than engaged. Yet it is precisely in that discomfort that many of the most valuable forms of understanding emerge. If those moments are repeatedly bypassed, the result is not a lack of output, but a reduction in depth. The leader may continue to produce answers, but with less direct contact with the process through which those answers are formed.

What is already clear is that the external world continues to accelerate, while the human system still depends on continuity, rhythm and recovery. Attention cannot be infinitely divided without consequence. It requires duration to stabilise, integrate and function effectively. When that duration disappears, fragmentation becomes normal, and when fragmentation becomes normal, leadership becomes thinner. More reactive. Less reflective. Less anchored in experience.

None of this is solved by going on a digital detox. We have tried. The detox ends, the phone comes back, and the pattern resumes within a week. A five-minute walk without the phone. Putting the device in another room during dinner. Reading three pages of a book before checking any screen in the morning. These sound trivial. They are not.

Reclaiming attention is not about rejecting technology. It is about restoring choice within the environment we already live in. It is about creating moments in which the mind is allowed to settle rather than constantly respond. These are not dramatic interventions. They are small, repeated decisions that gradually rebuild the capacity for presence. A meeting entered without a device. A conversation held without interruption. A task completed without switching. These moments may appear insignificant, but they begin to restore something fundamental. They rebuild continuity. They allow the system to stabilise. And from that stability, a different quality of leadership becomes possible.

Relmi has spent years building exactly this kind of practice into her own life. As someone who works at the speed her field demands, she has found that biohacking is not a luxury or a wellness trend, but a survival strategy. various types of water immersions, deliberate mind wandering, regular movement, sustained sleep and brief moments of closing the eyes throughout the day are, for her, the disciplines that allow attention to recover before it fragments beyond return. Her experience points to something the rest of us are still learning. In a world that pulls attention apart by default, the ability to gather it back has to be cultivated deliberately, and that cultivation begins in the body before it reaches the mind.

As hard as it is to accept, our attention has been colonised. It has been pulled apart, monetised and trained away from depth. The question now is no longer whether this is happening, but what we are willing to do about it. How do we reclaim the quality of our attention, and what are we each prepared to change in order to protect what remains most human in us? This is not a sentimental question. It is a leadership question.

Leadership, at its core, is not only about what we do, but how fully we are there when we do it. In a world that keeps pulling attention outward and apart, the leader who can gather it inward and hold it steadily may become rarer, but more necessary than ever.

Reflection Question

Think about the last time you were with someone important to you, at home or at work, while part of your attention was somewhere else.

What might full attention have allowed you
to notice in that moment?

CHAPTER 7

THE DISCONNECTED HUMAN

WHEN CAPABILITY EXPANDS BUT WISDOM RECEDES

When attention fragments long enough, the impact is no longer only cognitive. It begins to reshape how we experience ourselves. What starts as a shift in focus gradually becomes a shift in identity, altering not only how we think, but how we relate to our own thinking. We remain active, informed and capable, yet less anchored in our own internal processes. Answers arrive faster, but clarity does not deepen. Output increases, but inner authority weakens. There is movement, but less grounding within it. The external world accelerates while the internal system struggles to maintain coherence. We are not becoming less intelligent. We are becoming less connected to the processes that make intelligence meaningful. And it is within that disconnection that a subtler transformation begins to unfold, one that is not immediately visible but profoundly consequential.

When capability expands, it does not only change what we can do. It changes how we relate to doing itself. The surface of modern life suggests progress in its most visible form. Systems respond faster. Information is immediate. Decisions can be supported, generated and executed with increasing efficiency. From the outside, this appears as an expansion of human potential. Yet beneath that expansion, a redistribution is taking place.

As external systems begin to carry more of the cognitive load, the internal processes that once supported thinking are exercised less frequently. Reflection shortens. Discernment is bypassed. The time between question and answer compresses. The discomfort that often accompanies deeper reasoning, the pauses, the friction, the slow formation of insight, becomes something to avoid rather than something to enter. Over time, this reshapes the relationship between the individual and their own thinking. The question is no longer something we stay with. It becomes something we move through. And gradually, without noticing, the internal architecture that supports judgement begins to weaken.

This is where wisdom begins to erode. Wisdom is not a function of speed or access to information. It is the capacity to remain with complexity long enough for understanding to form. It requires attention that can hold tension without immediately resolving it. It depends on the ability to integrate multiple perspectives, consider context and sense consequences beyond what is immediately visible. When answers arrive instantly, they create the feeling of knowing without the process that gives knowing its weight. A response may be coherent, structured and convincing, yet still lack depth. It may sound complete without being integrated into lived experience. It may appear certain without having been tested through reflection. The individual continues to function, but with less contact with how their thinking is formed.

Inner authority erodes in the same way. Inner authority is not certainty, nor independence from others. It is the capacity to remain connected to one's own evaluative process while engaging with external input. It allows a person to question, to sense alignment or misalignment and to remain present with ambiguity without immediately resolving it. In an environment where answers are continuously generated and delivered, this capacity can weaken. The distinction between consulting a system and surrendering to it becomes increasingly subtle. A person may begin by using technology to support their thinking, but gradually rely on it to replace parts of that thinking. The presence of an immediate answer reduces the perceived need to remain with uncertainty. Reflection is replaced by retrieval.

A leader sits down to prepare for a difficult conversation. In another time, they would have had to think their way into it. What is really happening here. What is this person likely carrying. What am I trying not to see. What needs to be said, and what needs to be understood before I say it. Now the process can be shortened. A system produces a clear, balanced response in seconds. The structure is sound. The language is polished. It says all the right things. But something subtle has not happened. The leader has prepared the words without fully forming the mind behind them.

We can tell the difference now when we watch a leader work with AI. Some of them stay awake inside the process. They read what the system gives them and argue with it. They reject half of it. They take one phrase and build something the system could not have given them. Others accept what appears on the screen as if it were already theirs. The difference is visible in seconds, and it predicts almost everything about what happens next.

In this relationship, the system extends thinking without replacing it. To outsource judgement is something else entirely. It is to accept outputs because they are fluent, defer to structure because it appears complete, and move from question to answer without engaging in the internal work that would make that answer meaningful. The risk is not that we use intelligent systems. The risk is that we stop using ourselves in the process.

This does not create immediate failure. It creates a softer form of disconnection. A person may still perform, still deliver, still communicate effectively. Yet they do so with less depth, less integration and less connection to their own internal signals. This is what we have come to call the disconnected human: not uninformed, not incapable, but less connected to the processes that allow intelligence to mature into wisdom.

This disconnection is not only cognitive. It is embodied. The body participates continuously in perception, evaluation and response. Subtle shifts in sensation, tension and emotional tone provide early signals about alignment, risk and meaning. These signals often precede conscious thought. When individuals become less attuned to them, a significant layer of intelligence becomes less accessible. Leaders often recognise this in hindsight. They recall moments where something felt off, yet they moved forward because the analysis appeared sound. They sense that something was missing, but cannot immediately name it. The decision may succeed. The outcome may be positive. Yet a layer of coherence is absent. The result may work. The person is still left untouched by how they arrived there.

Over time, this creates distance. Decisions remain effective, but less integrated. Action continues, but with reduced depth. The individual remains capable, but less whole in how that capability is expressed.

Across different environments and cultures, a consistent pattern becomes visible. People are rarely limited by a lack of information. They are limited by the frameworks through which they interpret their experience, and by the degree to which they remain connected to themselves while moving through change. Identity is not fixed. It reorganises in response to context, challenge and experience. Each transition introduces a threshold. Each threshold requires a reconfiguration of how one understands oneself.

When individuals adapt externally without transforming internally, a misalignment begins to form. They update behaviour, but not perception. They respond to new demands without reconstructing the frameworks through which they interpret them. This is where knowledge accumulates, but does not transform. There is a difference between accumulating knowledge and being shaped by it. One expands what we know. The other reshapes who we are. Information can increase without the person deepening.

In a world where change is constant, this distinction becomes essential. The ability to question inherited assumptions, reconstruct identity and remain connected to one's own evaluative process is no longer optional. It becomes the foundation of meaningful adaptation. If that foundation weakens, the implications extend beyond the individual, because the decisions we make, the systems we design and the environments we create are all shaped by the level of integration we bring to them.

And this is where the next question becomes unavoidable. If we are becoming less connected internally, what happens to the systems we are building externally?

Reflection Question

Think about a recent decision where you moved quickly because the answer was available.

If you had stayed with the question a little longer, what might have emerged that required more of you than the answer itself?

CHAPTER 8

WHEN INNOVATION FORGETS THE HUMAN

PROGRESS WITHOUT ORIENTATION

When individuals become fragmented and disconnected, the systems they build begin to reflect the same condition. What we create carries the imprint of how we think, how we attend, and how we relate. Technology does not emerge in isolation. It is shaped by the state of the human beings designing it, consciously or not. So the question is no longer only about technology. It is about the state of the human system behind it. It is about the quality of attention, the depth of reflection, and the level of awareness that informs what is being built.

There is a question that sits beneath much of what we are creating, yet it is rarely asked with the seriousness it deserves. What happens when the race for capability moves faster than our capacity for care?

For centuries, innovation has been associated with improvement. To build more, to solve faster, to extend human reach has been seen as evidence of progress. Each generation has inherited tools that made life easier, safer, or more efficient than before. But capability does not carry direction. Efficiency does not guarantee meaning. Progress, when left unquestioned, can continue moving long after it has lost sight of why it began.

We are now building systems that can generate, predict, and respond at a scale no individual can fully comprehend. Entire domains of work are being reshaped. Decisions are increasingly influenced by models that operate beyond direct human visibility. And yet, within that acceleration, something is at risk of being overlooked. The systems we create reflect the assumptions and limitations of the people designing them.

When those individuals operate with fragmented attention and weakened inner authority, the systems they build begin to carry the same pattern. Not intentionally, but inevitably.

This is already visible in how some of the most advanced systems are being developed. A number of leading AI laboratories have begun to publish written principles for the behaviour of their models, attempting to embed values into the way these systems respond, refuse, and reason. The fact that this kind of documentation has become necessary is itself a statement about where innovation has arrived. It means that the question is no longer only what these systems can do. It is what they should do, and who decides. It is an acknowledgment that capability without values is not a neutral starting point but a hazardous one. A system that can generate persuasive language, recognise patterns across enormous datasets, and make recommendations that influence human decisions cannot be released into the world without some answer to the question of what it should be for.

And yet, this introduces a deeper tension. These systems are being built at extraordinary speed, while the values that guide them are still being defined in real time. Those values are shaped by relatively small groups of individuals, often operating under pressure to move quickly. Values are not neutral. They are choices. They reflect perspectives, assumptions, and priorities that shape how systems interact with human beings. The people writing these principles are doing serious work, but they are doing it inside a competitive landscape that does not pause for reflection. The pressure to ship, to demonstrate capability, to keep up with rivals, runs against the kind of slow, collective deliberation that questions of this magnitude would normally require.

If we were willing to pause, not for a product cycle but for a longer horizon, and ask more fundamental questions about those values, the trajectory of innovation might look very different. It might prioritise not only capability, but dignity. Not only efficiency, but psychological wellbeing. Not only speed, but long-term human development.

The contrast becomes clearer when we look at what happens when capability advances without that kind of consideration. Social media offers the clearest contemporary example, and it is one most readers have lived through directly. The platforms that now shape how billions of people communicate, consume information, form opinions, and relate to themselves were designed with extraordinary technical sophistication and very little reflection on their consequences for human attention, mental health, democratic discourse, or the developing brains of children. They were optimised for engagement, because engagement was what could be measured and what produced revenue. The values were not absent. They were embedded in the design choices, but they were narrow values, focused on what the systems could capture and hold rather than on what kind of humans would emerge from spending years inside them.

The consequences are now visible. Adolescent mental health has deteriorated in ways that closely track the introduction of smartphones and social media into the lives of young people (Haidt, 2024). Rates of anxiety, depression, and self-harm in teenagers have risen sharply over the past decade and a half, particularly among girls. Public discourse has become more polarised, more reactive, more driven by outrage than by understanding. The conditions for sustained attention have been eroded across a generation. None of this was the intended outcome of social media design. But it was the outcome, and it became visible only after the systems were already deeply integrated into everyday life.

The lesson is not that the engineers and designers were careless. Many of them are now among the most thoughtful critics of what they helped to build. The lesson is that capability moved faster than the human systems trying to evaluate it, and by the time the consequences became clear, the platforms had become infrastructure. They could no longer be paused or reconsidered without enormous cost. The window in which questions about values could have shaped the design had closed.

This is what is at stake now with artificial intelligence, on a much larger scale. The companies building these systems are aware of the risks in ways that the early social media companies were not, and the work being done on alignment, safety, and the articulation of guiding principles is genuinely serious. But the underlying pressure is the same. The race is on. The systems are being deployed before their long-term consequences can be understood. The values being built into them are being decided by a small number of people working at extraordinary speed. UNESCO's Recommendation on the Ethics of Artificial Intelligence has placed human dignity, rights, and wellbeing at the centre of how this development should proceed, emphasising the importance of human oversight and long-term societal benefit (UNESCO, 2021). Whether that framework translates into practice depends on whether the people building the systems, and the leaders adopting them, are willing to slow down long enough to ask the questions it raises.

This is not an argument against innovation. It is an argument for responsibility within it. Systems designed without consideration for human experience may scale quickly, but they do not sustain value. They optimise output, but reduce depth. They increase efficiency, but weaken the conditions that allow human beings to think, relate, and create meaning. Across years of working with leaders in organisations grappling with technology adoption, we have noticed a pattern. The conversations almost always begin with capability, what the new system can do, how much time it will save, what efficiencies it will produce. The conversations rarely include the question of what it will cost. Not financially, but humanly. What will it do to the rhythms of work, to the quality of attention, to the kind of thinking the team is able to sustain, to the relationships between people who will now be mediated by the system rather than meeting each other directly? These questions are not difficult to ask. They are simply not asked, because the framework within which the decision is being made does not include them.

This is where leadership becomes decisive. Leadership is no longer only about navigating complexity or delivering results. It is about shaping the conditions in which those results are produced. It is about deciding what kind of systems we are willing to build, and what kind of human experience those systems will create.

A leader designs a process to increase efficiency. Output improves. Metrics rise. But over time, something shifts. Conversations shorten. Reflection decreases. People stop bringing the kind of slow, careful thinking that the new process makes feel inefficient. Work becomes faster, but thinner. Nothing has gone wrong in a measurable sense. And yet something has been lost. This is how distortion enters. Not through failure, but through imbalance.

When capability advances without corresponding development in human awareness, the result is not progress. It is fragility. Systems that optimise for speed without depth become increasingly dependent on conditions they cannot sustain. They require constant input, constant output, and continuous acceleration. They lose the ability to pause, to reflect, and to recalibrate. Over time, they begin to break. Not because they are weak, but because they are incomplete.

This is the deeper risk of our time. Not that machines become more intelligent, but that humans begin to organise the world around what machines can do, rather than what human beings need. The reorganisation is already underway. Work is being redesigned around tasks AI can perform, with the human contribution shrinking to whatever the machine cannot yet handle. Education is being restructured around the assumption that students will use AI tools, with less attention to what kind of mind develops when essential cognitive struggles are bypassed. Creative industries are being rebuilt around generated content, with the slower, harder work of original thinking being priced out of the market. None of these shifts are inevitable. But they are happening, and they are happening because the question being asked is what is now possible, not what is now wise.

If thinking is increasingly supported externally, then the human frontier shifts. It moves toward what cannot be replicated: the capacity to feel, to imagine, to sense meaning, and to remain present in experience. This is not a retreat from progress. It is a reorientation of it. To move forward in this environment requires not only technological literacy, but human depth. It requires the ability to remain connected to attention, to reflection, and to embodied experience even as external systems accelerate. The question is no longer only what can be built. It is what must be preserved.

To preserve humanity within progress requires conscious choice. It requires leaders who are willing to pause, to question, and to remain connected to the human impact of what they are creating. The real risk is not that machines will replace us. It is that we begin to imitate them. And that is a choice that is being made every day, in every organisation, by people who often do not realise they are making it.

Reflection Question

Think about a system, process, or decision you are currently shaping, at work or in your broader life.

Where might efficiency or capability be taking priority over human experience, reflection, or connection?

PART III

THE FOUR HUMAN ABILITIES OF LEADERSHIP

THE HUMAN EDGE IN THE AGE OF AI

CHAPTER 9

BEHAVIOUR IS THE SURFACE

IF BEHAVIOUR IS WHAT WE SEE, THESE ARE THE FORCES THAT DECIDE WHAT BECOMES POSSIBLE

By this point in the book, one thing should be clear. Leadership does not begin where it becomes visible. What we see in a leader, their decisions, their communication, their presence, is not the origin. It is the surface, the final expression of processes already unfolding beneath awareness.

In Part I, we explored the human system beneath leadership and made the case that behaviour is not the starting point, but the expression of deeper biological, emotional and cognitive patterns. In Part II, we examined what happens when that system is placed inside a world of continuous acceleration, fractured attention, weakened inner authority and technological expansion. We looked at stress, dysregulation, fragmentation and the subtle ways modern life can make people more reactive, less present and less coherent than they appear on the surface. Taken together, those chapters lead to a necessary conclusion. If behaviour is not the origin of leadership, then leadership development cannot remain focused on behaviour alone. It must move beneath it, into the forces that shape whether clarity, steadiness, empathy, creativity and discernment are available in the first place.

This is where most leadership models begin to fail. They start at the surface. There is a reason for this. Behaviour is visible. It can be described, measured and trained. It creates the illusion of control. But behaviour is not where leadership begins. It is where leadership ends. Everything that matters has already happened before it appears.

When Silvia first introduced the i4 Neuroleader Model in Leadership Is Upside Down (Damiano, 2014), the argument was simple and, at the time, quite radical: leadership development was starting too late. It was focusing on what leaders should do without sufficiently understanding what allows a leader to do it. Even then, that was already a significant move. What she proposed instead was that four human abilities needed to be cultivated if leadership was to be sustainable, human and effective: Integration, Inspiration, Imagination and Intuition. These were not presented as abstract qualities. They were directly connected to the capacities leadership requires in practice: Performance, Collaboration, Innovation and Agility.

What could not yet be fully seen, when the model was first introduced, was how dramatically the world would intensify the need for it. A global pandemic would place nervous systems under prolonged uncertainty. Geopolitical instability would increase collective stress, fear and emotional fatigue. Artificial intelligence would begin to reshape the nature of work, cognition and value itself. The pressure leaders now face is not episodic. It is ambient. It is woven into the atmosphere of daily life. In that kind of environment, a behaviour-first model becomes even less sufficient. People do not fail because they lack advice. They fail because, under real conditions, they cannot always access what they already know. That is the real gap.

At the time the model was first developed, the scientific language around these ideas was still emerging. The model was built on a combination of early neuroscience and a deeper, intuitive recognition of how humans actually function under pressure. Since then, the evidence has expanded. Research has made clearer what was already being observed: stress narrows cognition, attention fragmentation reduces clarity, emotional states shape decision-making, and creativity and perception depend on conditions, not just capability.

What was once insight has become increasingly difficult to ignore. Still, the gap between what leaders know and what they can access remains.

This did not begin as theory. For Silvia, it began as a collapse.

"In 2008, during the global financial crisis, I was working as a leadership development specialist, helping other people navigate stress, uncertainty and pressure. From the outside, I looked functional. I was still delivering programmes, still supporting clients, still doing what I was supposed to do. But inside, something was changing, and I could not explain it through mindset or behaviour alone. Energy dropped. Focus narrowed. Emotions became less stable. Mornings grew heavy. Even getting out of bed became difficult.

That gap between how I looked and how I felt became impossible to ignore. When I finally sought medical support, a brain scan revealed something that changed the way I understood the whole experience. What was happening was not simply emotional. It was biological. Activity in parts of the brain linked to mood regulation was no longer in balance. What had felt like personal failure was, at least in part, a dysregulated system.

That moment changed the direction of my work. It made something unmistakably clear: the distance between how a person appears to be functioning and what is actually happening in their system can be enormous. And that gap cannot be closed by behaviour alone. Recovery did not come from trying harder. It came from learning to work with the system itself. Physiology. Attention. Environment. Movement. Breath. Rest. Emotional processing. Slowly, coherence began to return. Not perfectly. But enough to show that this was where the real work had to begin."

What followed was a four-year period of formal research in which Silvia drew together two decades of work in leadership development, executive coaching and consulting with leaders across the globe, and gave that experience the structure of a methodology and an assessment. The research combined qualitative and quantitative methods, statistical analysis, a literature review at the intersection of neuroscience and leadership, focus groups, interviews with leaders around the world and the lived experience of testing the practices on her own brain and body.

It drew on the strengths of the team Silvia had assembled around her, each member contributing a particular form of expertise. Relmi was present every step of the way, as the creative force whose work would translate the findings into a language others could see, recognise and enter. As Silvia distilled what she was learning into early books, white papers and assessments, Relmi built the brand, the visual identity and the communication architecture that would carry the model into the world.

What began as a personal turning point became a line of inquiry. How many leaders are operating effectively on the surface while their internal system is under strain? How often is behaviour being corrected when the real issue is biological, cognitive and emotional misalignment? What would leadership development look like if it started from there? The answer that emerged from those questions became the foundation of everything that followed. Leadership rarely breaks at the level of behaviour. It breaks earlier, at the level of perception.

Across years of working with the i4 Neuroleader Model, one pattern has appeared again and again. People do not experience themselves as others experience them. They often see themselves as less capable, less confident or less effective than they actually appear in their interactions.

Others, meanwhile, experience them as composed, thoughtful and capable of navigating complexity with more stability than they recognise in themselves. This gap matters, because behaviour does not respond only to reality. It responds to interpretation. If a leader believes they are not ready, not strong enough or not clear enough, their behaviour will reflect that internal state regardless of their actual capability. Leadership is not limited only by skill. It is limited by the accuracy of self-perception.

This is not merely a coaching observation. A wider body of work points in the same direction. David Rock has shown how threat and overload narrow leadership capacity. Carol Dweck has demonstrated how beliefs shape growth. Tomas Chamorro-Premuzic has argued that confidence and competence are often far less aligned than people assume (Rock, 2008; Dweck, 2006; Chamorro-Premuzic, 2019). The language varies across their work. The pattern does not. The system beneath behaviour determines what becomes possible.

This is why the i4 Neuroleader Model does not begin with behaviour. It begins with awareness of the system producing that behaviour. The model recognises that leadership is shaped by the interaction between biology, attention, emotion and cognition long before any action becomes visible. What we see externally is the final expression of patterns already in motion internally. Across assessments, debriefs and practitioner experience, one insight repeats itself: people are not simply lacking capability. They are disconnected from it. They underestimate their strengths, overidentify with their perceived limitations and operate from internal narratives that compress their ability to act. This is not a performance issue. It is a coherence issue. And coherence cannot be built by trying harder at the level of behaviour. It has to be developed at the level where behaviour is formed.

This is why the model stands a chance now, perhaps more than when it was first conceived. Not because it is elegantly named. Not because it offers another branded answer in an already crowded field. It stands a chance because the world we have described in the previous parts of this book is now demanding, with increasing urgency, the very capacities it names. Part I showed that leadership does not begin with behaviour, but with the condition of the human system. Part II showed what happens when that system is exposed to chronic pressure, fractured attention, weakened inner authority and the accelerating force of technology. If the book ended there, it would offer a sharp diagnosis. But diagnosis is not enough. Part III must now name the response. It must make visible the deeper human capacities that need to be cultivated if leadership is to remain human, adaptive and wise. That is where the i4 Neuroleader Model comes fully into its relevance. Not as a framework laid over the material from the outside, but as the natural consequence of everything the book has already revealed.

The four human abilities at the heart of the model are not powerful because they are clever. They are powerful because they have become necessary. As external intelligence expands, as cognition is increasingly supported, accelerated and outsourced, what rises in value is not more information, more process or more polished competence. What rises is the ability to remain whole under pressure, generate meaning and energy without coercion, imagine what does not yet exist and perceive beyond what is immediately visible.

That is *Integration*, *Inspiration*, *Imagination* and *Intuition*.

The model matters now because it is not trying to compete with machines on machine territory. It points instead toward the distinctly human capacities that must be protected and developed more deliberately as the world becomes faster, louder and more artificial.

There is another reason this work has continued to evolve rather than remain fixed. From the beginning, it was never only a body of ideas. It was something that needed to be brought into lived experience. Relmi has carried that expression from the start. Her contribution has been creative, relational and generational, formed within a generation that inherited earlier leadership theories while living inside a radically different cognitive environment. From her perspective, the model has never been only about professional effectiveness. It has always pointed to something larger. How does a human being remain connected, creative and fully present when the boundaries between work and life are dissolving, attention is under siege and thinking itself is becoming easier to bypass? That question gives the model a different kind of urgency. What began as a leadership methodology now speaks to something more fundamental. These are not only the capacities required to lead well. They are the capacities required to live well, think deeply and remain in genuine connection with others while the conditions of modern life keep pulling us toward fragmentation.

This is also why the model has resonated so deeply with practitioners. In a public conversation about their work with the model, two American practitioners, Courtney McBath and Emily Juricek, described what drew them to the i4 Neuroleader™ approach and what made it different from other leadership frameworks they had encountered. What stands out in their reflections is not promotional enthusiasm, but practical recognition.

They return repeatedly to the same point. Most leadership models deal with symptoms. This one helps them work with causes. That distinction is enormous. In coaching, the symptom might be low innovation, weak trust, difficulty with regulation, chronic burnout or hesitation in decision-making. Traditional leadership development often responds by prescribing better behaviours. Speak differently. Encourage more. Become more strategic. Be more resilient.

But as Courtney observed, that approach does not always go far enough, because people may already be self-aware and still unable to change. Emily made the same point in different language. People do not only want to know how they are. They want to know how to shift. That is where the methodology becomes meaningful. It does not stop at insight. It offers structure, resources and a practical pathway.

Self-awareness alone is not enough. This is one of the most important truths in leadership development, and one of the most misunderstood. Many people assume that once they can see the problem, change will naturally follow. But awareness without structure often leads to frustration. People become more conscious of their patterns and yet continue reproducing them. They know they are reactive, depleted, hesitant or overextended, but the knowing itself does not reorganise the system. The practitioners in that conversation are blunt about this. Self-awareness did not fix them. What helped was understanding how to work with the brain, body and behaviour together. That is a much stronger proposition. It moves leadership development out of vague aspiration and into practical neuroplasticity. Not simply think differently and hope your life follows, but think differently and act differently so the underlying pathways begin to change.

This is one of the great strengths of the i4 Neuroleader™ Methodology. It was never designed to be a static diagnostic tool. It was designed as a developmental ecosystem. The assessment is only the starting point. From there, the person is led into a broader process through articles, courses, books, white papers, coaching prompts and practical next steps intended to help them strengthen the areas highlighted in the report. The model does something many leadership tools do not. It does not merely show people where they are. It gives them a way to move.

This is also where the work becomes practical, not in theory but in application. Through decades of working with leaders and organisations across Europe and Latin America, Juan Carlos has seen this pattern repeat itself in every context where performance matters and pressure is constant. What emerges in those environments is not complexity, but clarity. Leaders who understand what is required, yet cannot sustain it. Leaders who know how to collaborate, yet default to control under stress. Leaders who are capable, yet inconsistent. The issue is rarely knowledge. It is access. And access is a function of the system.

At one level, the model was built to develop four major competencies essential for life and work: *Performance*, *Collaboration*, *Innovation* and *Agility*. Each one matters.

Performance matters because leaders need mental and physical capacity to function well. *Collaboration* matters because leadership is relational. *Innovation* matters because the future does not reward repetition alone. *Agility* matters because conditions keep changing. The supporting pillars beneath these competencies matter too and, in practice, they are useful, measurable and often highly revealing.

But for the purpose of this book, and for the argument we are building, we focus on the four human abilities that sit at the core of the model: Integration, Inspiration, Imagination and Intuition. Not because the remaining abilities are unimportant, but because these four offer especially powerful leverage points for understanding what makes a human being able to function, relate, create and decide well in a world moving too fast, demanding too much and increasingly externalising intelligence. They also need to be kept distinct. If they blur into one another, the model weakens. If they are understood clearly, the model lands.

Integration is the capacity to remain whole. It is the alignment of brain, body and mind that allows a leader to stay present and stable when pressure rises. Without it, the other abilities become unstable.

Inspiration is the capacity to generate energy, meaning and direction without coercion. It is what allows a leader to move others through aliveness rather than through pressure.

Imagination is the capacity to see what does not yet exist. It is the ability to think in possibilities rather than only in procedures, and to construct futures that are not extensions of the present.

Intuition is the capacity to perceive before explanation. It is a form of knowing that arises from embodied experience and pattern recognition, before analysis has caught up.

Together, these four abilities offer a map of deeper leadership development. They help explain why some people remain steady under pressure while others fragment. Why some can generate energy and possibility while others remain trapped in effort and repetition. Why some know how to pause and trust what they sense while others drown in analysis. These are not decorative qualities. They are decisive.

This is also why the model has become more relevant in the current era, not less. As the world becomes more technologically capable, more data rich and more accelerated, the qualities that rise in importance are not merely technical. They are human. Artificial intelligence can help leaders write faster, respond faster and produce more. But if those gains are simply poured back into more output, without more reflection, more rest, more clarity or more humanity, then capability becomes a trap. That is precisely why the internal work cannot be optional. It has to become part of leadership itself.

What follows in the next chapters is not a model layered on top of leadership. It is a return to the human system that makes leadership possible. These chapters are not about leadership as performance theatre, but about leadership as human development. They also reflect a truth that has become clearer across generations: what was once required to lead well is now increasingly required to live well.

We explore the four human abilities not as abstract concepts, but as living capacities that can be strengthened. We begin with Integration, because without coherence the rest quickly becomes unstable. Then we move into Inspiration, Imagination and Intuition, each opening a different dimension of what it means to lead in a deeply human way.

If behaviour is the surface, and if the modern world is fragmenting the system beneath it, then this is where we begin. Not with what leaders do, but with what allows them to do it.

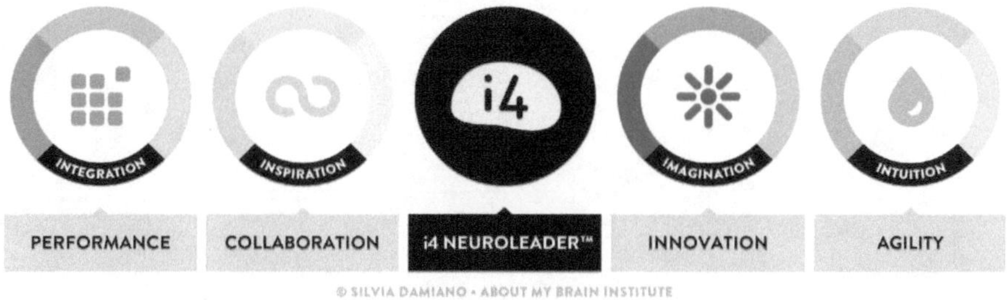

To view the coloured version of the model and learn more about the methodology visit:

www.aboutmybrain.com/i4model

“

**YOU DON’T LEAD FROM WHAT YOU KNOW.
YOU LEAD FROM WHAT YOU CAN ACCESS.**

”

CHAPTER 10

INTEGRATION
THE CAPACITY TO BE WHOLE

There is a version of leadership that looks competent from the outside yet feels unstable from within. Decisions are made, meetings are led, strategies are executed, and to everyone watching, the person appears capable enough. Yet beneath that surface there is strain. Attention is scattered. Emotions fluctuate more quickly than usual. Energy comes in spikes rather than in a steady current. Under pressure, something shifts. Clarity narrows. Patience shortens. The leader who seemed composed an hour earlier becomes less available, less perceptive, less themselves. This is not a failure of knowledge. It is a failure of integration.

One of the strongest insights emerging from practitioners working with the i4 Neuroleader Model is that many leadership difficulties are not behavioural failures, but regulation failures. People often know how they want to lead. They understand the principles. They can articulate the values. They may even coach others beautifully. Yet in the moment that matters, when pressure rises, timelines tighten or the emotional temperature in the room changes, they lose access to the very capacities they intended to use. What disappears is not intelligence itself, but coherence: the ability of the system to remain sufficiently aligned for perception, emotion and action to work together rather than against one another.

Integration is the capacity to remain whole under those conditions. It is the ability to bring brain, body and mind into enough coherence that the leader can stay present and stable in the face of complexity, not because complexity disappears, but because the internal system is less easily pulled apart by it. Without it, leadership becomes dependent on circumstances.

A person functions well when conditions are favourable and struggles when they are not. With it, that relationship begins to reverse. The leader becomes less easily hijacked by stress, more able to recover from disruption and more available to others in real time.

What often appears as inconsistency in leadership is not a lack of discipline. It is a lack of internal coherence. Many individuals can think clearly in calm conditions, yet lose access to that clarity under pressure. The issue is not knowledge. It is the ability of the system to remain regulated when demands increase. When the brain, body and emotional system fall out of sync, reaction replaces response. A person may still sound intelligent, still keep moving, still produce output, but the quality of their presence changes. Their behaviour becomes more situational, more brittle and less trustworthy over time. Fragmentation does not announce itself. It accumulates bit by bit, until the moment you most need coherence is the moment it is least available.

Integration matters so much because it is the foundation beneath all the other human capacities explored in this part of the book. Inspiration without integration becomes unstable. Imagination without integration becomes scattered. Intuition without integration becomes unreliable. Before a leader can energise others, create what does not yet exist or sense what is emerging, they need access to themselves.

In Chapter 3, we began to speak about the integrated self, not as a poetic idea but as a biological reality. The head, heart and gut are not isolated centres. They are part of one living system, communicating continuously through neural, hormonal and electrical pathways. Within the brain itself, integration is equally essential. Executive function, emotional processing, memory, bodily awareness and attentional control do not operate as separate systems. They depend on coordination. Leadership depends on their ability to work in unison.

This is not only an English-language conversation. Some of the most interesting work happening right now is in Spain, where a new generation of neuroscientists is asking exactly the questions this book is asking, and asking them within a different philosophical tradition that brings different textures to the answers.

Dr Nazareth Castellanos, whose research focuses on the relationship between brain activity and cardiac, gastric and respiratory dynamics, argues that the mind cannot be understood through the brain alone. In her recent work, she explores the gut-brain, heart-brain and breathing-brain axes, presenting breathing as a bridge between the exterior and interior worlds (Castellanos, 2025). That matters for leadership because it reinforces a point too often forgotten: the quality of our thinking depends, in part, on the state of the body doing the thinking.

Integration, then, is not about control. It is about alignment. The same principle becomes visible through creativity. The question is not how to become more original, but how to remove the interference that blocks access to what is already there. Creativity, like clarity, does not emerge well from a fragmented system. When attention is scattered, the body is overstimulated and the inner environment noisy, expression becomes forced. But when the system settles, something opens. Thought becomes less effortful. Perception sharpens. What appears on the surface as creativity is often a reflection of deeper coherence. In that sense, integration is not only the foundation of leadership. It is also the foundation of access: to thought, to feeling, to insight and to the parts of ourselves that pressure usually suppresses.

This distinction matters because much of what passes for self-management in leadership is little more than suppression. A person learns to contain emotion, regulate tone and maintain the appearance of calm. But the appearance of calm is not the same as internal steadiness. A leader can continue functioning while privately dysregulated. Many do. The problem is that fragmentation always leaks. It shows up in impatience, impulsive decisions, erratic energy, poor sleep, reduced empathy, difficulty staying focused and the slow erosion of resilience. The more useful question is not whether performance is still possible, but how integrated the system producing it actually is.

This is where the neuroscience becomes useful. Mark D'Esposito's work on executive function and working memory shows how much leadership depends on the prefrontal cortex's ability to hold information in mind, filter distractions, regulate impulses and adapt behaviour in real time (D'Esposito & Postle, 2015). These are structural capacities. They determine whether a person can maintain focus, shift perspective and think strategically when the environment becomes demanding. Stress, fatigue and overload impair exactly these functions.

The issue is not merely that people feel worse. The issue is that the very neural systems required for wise leadership become less available. Under pressure, the brain defaults to efficiency. It conserves energy by reaching for familiar patterns and narrowing its field of view to the most immediate demands. This is useful in genuine emergencies but corrosive when it becomes habitual. A leader with compromised executive function may struggle to prioritise, lose sight of the larger arc, become more distractible and react more quickly than they intend. What appears as impatience or inconsistency is often a system that has lost coherence.

Research on high performers points in the same direction. Dr Fred Travis and Dr Harald Harung found that individuals operating at very high levels across business, sport and leadership showed greater brain integration and coherence than comparison groups (Travis & Harung, 2015). We have seen this in leaders who should not, by any conventional measure, be performing at the level they are. They are not the smartest people in the room. They are not the most credentialed. What they have instead is a kind of internal silence that lets them think clearly when everyone else is narrowing. This silence is not a personality trait. It is the signal of a system that is actually working.

Work on contemplative practice suggests that when internal noise decreases, clarity and cognitive stability increase (Rosenthal, 2016). Studies of peak performance describe states in which the usual fragmentation of self-consciousness quiets and the system operates with greater fluidity and coordination (Kotler & Wheal, 2017). What matters is not the pursuit of extraordinary states, but what they reveal. When inner friction reduces, coherence becomes available.

In the language of this book, integration is not about chasing extraordinary states. It is about making coherence more available in ordinary ones. Can you remain clear in a difficult conversation? Can you stay present when tension rises? Can you think well when you are tired, uncertain or under pressure? This is where integration stops being an idea and becomes a lived capacity.

There is also a relational dimension that leadership literature often underestimates. Castellanos' work reminds us that breathing, attention and bodily state are not only personal processes. They shape how we meet others. If breathing can act as a bridge between inner and outer worlds, then regulation is not a private achievement. It affects the field around us. A regulated leader does not merely feel better. They alter the conditions in which other people think, respond and contribute. Presence is not charisma. It is not polish. It is the quality of being sufficiently integrated that others do not have to spend unnecessary energy defending themselves against your instability.

Many leadership problems are misdiagnosed because of this. Continuous connectivity, constant inputs and the speed of modern work increase the risk of fragmentation. The body is asked to absorb what the mind frames as normal.

The nervous system carries the cost. What looks like poor communication may be exhausted executive function. What looks like irritability may be chronic overstimulation. What looks like disengagement may be a system that has lost the energy required for regulation. None of this removes responsibility. It improves accuracy, and accuracy changes the intervention. If the issue is treated only as behaviour, the person is asked to try harder. Speak more carefully. Be more strategic. Stay calm. Listen better. These instructions are not wrong. They are simply too late. If the issue is understood as integration, the person can begin to strengthen the system beneath the behaviour.

Between us, we have coached hundreds of people across many years, industries and countries. And one pattern repeats more than any other. At some point, with hesitation, people admit a lack of confidence. They question whether the role is too big for them. They describe what is now widely called impostor syndrome. Or they acknowledge that their workload, environment or surrounding conditions are affecting their wellbeing in ways they have not been able to say out loud, because saying it has felt like admitting weakness.

The pandemic shifted some of this. It made it briefly more acceptable to acknowledge limits. But across all these conversations, one truth remains constant. People's mental and physical state shapes what they are able to do. Not occasionally. Continuously. This is what makes the biological perspective so useful. Once a person understands how sleep, stress, attention, emotion and physiology shape leadership behaviour, change becomes more grounded and more sustainable. The question shifts from "Why can't I be better?" to "What conditions make my better functioning possible?"

Gradually, coherence returns. Not perfectly, but sufficiently. What changes is not the complexity of the world, but the stability of the system meeting it.

This is one of the strongest contributions of the i4 Neuroleader Model. It does not ask leaders to optimise isolated traits. It asks whether the person is functioning as an integrated whole. That question is becoming more urgent in the age of artificial intelligence. As machines absorb more analytical tasks, more pattern processing and more cognitive output, the distinctly human advantage does not lie in becoming faster or more machine-like. It lies in becoming more coherent. Leaders will increasingly need to bring what machines do not: presence, regulation, ethical discernment, relational steadiness and the capacity to remain human under pressure. None of that is possible without integration. The future of leadership will not be defined by how much you can produce, but by how much of yourself you can reliably access under pressure.

Within the i4 Neuroleader Model, Integration is not one pillar among many. It is the ground on which the rest stand. When integration is weak, performance becomes erratic, collaboration grows strained, innovation turns forced and agility becomes reactive. When integration is strengthened, leadership becomes more stable, more sustainable and more trustworthy. Integration does not make the world less complex. It makes the leader less easily fragmented by complexity. It does not eliminate challenge. It changes the quality of response available in the middle of it. It allows the leader to stay with reality long enough to meet it well. And that is the beginning of brain-friendly leadership.

Integration is not something others can see directly, yet its effects are always visible. It shapes how you respond under pressure, how quickly you recover from disruption and how consistently you can access your own capability when conditions are least forgiving. The question is not whether you know how to lead. The question is whether your system allows you to.

Reflection Question

Think of a recent moment when you did not show up
as the leader you intended to be.

Was it truly a lack of knowledge,
or was it a loss of internal coherence?

What would change if, instead of correcting the behaviour first,
you strengthened the system beneath it?

CHAPTER 11

INSPIRATION

WHY PEOPLE CHOOSE TO FOLLOW YOU

Some leaders have everything in place and still cannot move anyone. The title is correct. The strategy is sound. The meetings are running. The metrics are tracked. From the outside, leadership appears to be happening. From inside the team, nothing is actually shifting. People do what they are asked, but not more. They show up, but they do not lean in. They comply, but they do not commit. What is missing is not the right action. It is the state from which action is taken.

For decades, leadership development has focused on what leaders should do: the behaviours, the techniques, the frameworks for communication and influence. These matter. But they do not explain the gap between leaders who are followed in name and leaders who are followed in spirit. Two people can deliver the same message, in the same room, with the same words, and produce entirely different responses. One is heard. The other is felt. One produces movement. The other produces silence dressed up as agreement. What separates them is not technique. It is state.

Inspiration, in the sense we mean it here, is not motivational performance. It is not the ability to deliver a stirring speech or summon enthusiasm on demand. It is the capacity to generate energy, meaning and direction without coercion, and to transmit that state in a way that activates something in others. It is what allows a leader to expand the field around them rather than merely occupy it.

When inspiration was first introduced as one of the four core human abilities of the i4 Neuroleader Model, it did not fit comfortably into traditional leadership categories. It seemed too abstract to measure, too internal to teach, too unlike the other competencies leadership programmes emphasised. It was included anyway, because the pattern was undeniable. The leaders who created lasting impact were not just efficient or strategic. They were alive in a particular way. They brought energy into rooms rather than draining it. They expanded thinking rather than narrowing it. They made people feel that something worth doing was being done.

A leader cannot hope to inspire others without first having access to something inspiring within themselves. This is one of the clearest insights from years of working with the model. Inspiration begins not in communication, but in internal activation. Before a leader can move others, something has to be alive in them. That aliveness is not a matter of personality or extroversion. Quiet leaders inspire as powerfully as charismatic ones, sometimes more so. What matters is whether the internal source is connected. When it is, people sense it. When it is not, no amount of technique can disguise the absence.

The neuroscience helps explain why. The human brain is not a neutral receiver of information. It is constantly evaluating its environment for signals about safety, meaning and effort. When a leader speaks, the listener's brain is not just processing the words. It is reading tone, emotional state, intention and congruence between what is being said and what is being expressed underneath it. If the signals align, something opens. If they do not, something protects itself. This happens beneath awareness, faster than thought, and it shapes everything that follows: whether a person engages or withdraws, whether they trust or hold back, whether they bring their full thinking to the work or only a managed version of it.

This is why some leaders speak and nothing moves, while others say very little and everything shifts. The difference is not what they say. It is what their nervous system is transmitting while they say it. Leadership, in this sense, is less about message and more about field. The leader is the field shaper, whether they intend to be or not.

This brings us to one of the most underappreciated aspects of leadership: emotional contagion. The phenomenon is real and well documented. The brain automatically synchronises with the emotional tone of others through mechanisms that include mirror neuron systems and rapid affective processing (Wicker et al., 2003; Keysers & Gazzola, 2014). We do not just observe other people's states. We internally simulate them. When a leader walks into a room carrying tension, the room tightens before a word is spoken. When a leader walks in carrying clarity and presence, the room settles. This is not motivational theory. It is biology. The leader's state is not a private matter. It is transmitted continuously, through facial expression, voice, posture and the subtle cues the body cannot fully hide.

The implication is uncomfortable. If you are anxious, you will spread anxiety. If you are disengaged, you will spread disengagement. If you are inspired, you will spread the conditions in which others can be inspired too. There is no neutral state from which leadership is exercised. Every interaction is a transmission. Every meeting changes the field. Over time, those transmissions become culture. Self-awareness in leadership is not optional, then. Before every important interaction, there is a question worth asking honestly: what state am I about to bring into this room? Whatever the answer is, it will reach the people in front of you, regardless of what you say.

Juan Carlos has long argued that inspiration begins where vocation is discovered. Drawing on the Spanish baroque writer Baltasar Gracián, he describes vocation as the discovery of one's own seed, the particular form of greatness each person carries and that can only be expressed through them. Vocation, in this framing, comes from voice, from being called. It is what allows a person to bring their full energy to the work they do, because the work itself reflects who they are. Without it, leadership becomes performance, and performance is exhausting in a way that vocation never is.

He pairs this insight with the psychological concept of fluidity, the state in which challenge and capability meet, where a person is neither bored by what they are doing nor overwhelmed by it. Across three decades of executive coaching, he has watched leaders who have not yet found that meeting point struggle to inspire others, no matter how strong their technical skills, while leaders who have found it generate the kind of energy that does not need to be manufactured. Inspiration, for him, is not a quality the leader brings to the work. It is what arises when the leader is in the right relationship with the work itself.

For years, leaders were taught the opposite. Stay composed. Stay neutral. Control your emotions. Do not let anything show. The intention behind this advice was reasonable: to prevent leaders from offloading their volatility onto their teams. But it produced a different problem. It created leaders who were emotionally flat, and flat leadership is not safe leadership. It is unreadable leadership. When a leader shows nothing, the people around them have to guess, and the brain interprets ambiguity as risk. What looks like composure from the outside often produces wariness on the inside. The leaders who inspire are not emotionally flat. They are emotionally real. They show enthusiasm when something matters, concern when something is wrong, optimism when the path is unclear. They do not perform these emotions. They allow them, with discernment, in service of clarity rather than catharsis. This is not weakness. It is signal. It tells the people around them that a human being is in the room, that what is happening is being registered, and that the leader is responding to reality rather than managing an image.

From years of working with organisations, we have observed the same pattern again and again. The people who genuinely inspire others share a small set of habits that are easy to overlook because they do not look like leadership techniques.

They take care of their physical energy, knowing that depletion silently undermines everything else they try to do.

They carry an underlying optimism, not as performance, but as a stance toward what is possible. They bring humour into rooms that would otherwise feel heavy. And above all, they are genuinely interested in the people around them, in their lives, their moments and their thinking, not as a relational tactic, but because they actually care. These are not glamorous qualities. They are subtle ones. But they are the conditions from which inspiration arises, and they are almost always present in the leaders others choose to follow.

What follows is one of Silvia's clearest lived examples of that pattern.

"Some of the best leaders I have worked with are not the ones who simply endure pressure, but the ones who organise their lives so pressure does not take over the system. One CEO I worked with over several years, described consistently by others as exceptionally calm, optimistic and demanding in the best sense, protected his energy with unusual discipline. Before the day began, he walked by the river, moved his body and entered work already regulated. What stood out to me was not only his stamina, but the effect of his steadiness on everyone around him. He did not lead through force. He led through coherence."

There is another place where inspiration becomes visible, and it is often more intimate than leadership literature allows. It appears in conversation. Not as communication in the functional sense, but as the moment where people come back into contact with themselves and with each other. In an age saturated with content yet increasingly starved of depth, genuine dialogue begins to carry unusual power. A real conversation slows the mind, organises thought and restores energy in a way that information alone cannot. It is one of the few spaces left in which people feel fully met rather than managed.

This understanding did not remain theoretical for us. It became something we wanted to build into lived experience, and it found expression in one of the products our family created through About my Brain®.

This insight sits beneath the creation of the Heart2Heart Conversation Cards. They were not designed as prompts to fill silence, but as invitations into the kind of exchange many people miss and rarely know how to re-enter. Their power lies in their simplicity. A question held with sincerity can open depth, dissolve performance and bring aliveness back into the room. In that sense, inspiration is not only something a leader generates internally. It is also something that can arise between people when the conditions for real presence are restored.

The stories we tell about leadership, whether drawn from film or from history, return to the same archetype again and again. Think of William Wallace in Braveheart, riding along the front line at Stirling, raising his voice above the fear of his men, refusing the option of retreat. Maximus in Gladiator, fighting in the front rank rather than behind it, knowing the names of the men around him, reminding them that what they did in life would echo in eternity. And then there is Ernest Shackleton, whose story is not fiction at all. When his ship, the Endurance, was crushed in the Antarctic ice in 1915, he led twenty-seven men through nearly two years of impossible conditions and brought every one of them home alive. He walked among his men daily. He gave away his own gloves when one of the crew lost theirs. He made sure he was the last to eat. He insisted on humour, routine and small celebrations even when survival itself was uncertain. He understood, without ever using the language of neuroscience, that the state of the leader becomes the state of the group (Lansing, 1959).

These figures do different things in different eras, two of them imagined and one of them real. But the signal is the same. The inspirational leader is present, physically and emotionally.

They are with their people rather than above them. They speak louder when fear is loudest. They show enthusiasm for what matters. They refuse to accept defeat as the first option. Above all, they go first. They are willing to step into the discomfort they are asking others to step into.

We tell these stories because something in us recognises them as true. The leader who is willing to be present, to be human, to carry weight visibly rather than hide it, is the leader people will follow into difficulty. The corporate version looks completely different. There are no battles, no speeches before combat, no Antarctic crossings. But the underlying signal is identical. The leader who walks the office floor asking how someone's weekend was, who acknowledges that a project is hard before asking the team to push further, who shows up to the difficult conversation rather than delegating it, is doing in a modern register what Shackleton was doing on the ice. The act looks completely different. The transmission is the same.

Teams do not trust perfection. They trust authenticity with direction. A leader who can acknowledge difficulty while still pointing toward what comes next is more trusted than a leader who pretends everything is fine. The first communicates that reality is being faced. The second communicates that something is being hidden, and the brain treats hidden things as threats.

Inspiration is fundamentally different from motivation. Most leaders, when they want to drive performance, push. They use deadlines, targets, pressure and consequences. Sometimes this works in the short term, because pressure activates the survival systems of the brain and forces action. But survival systems are not creative systems. They produce compliance, not contribution. They narrow attention, not expand it. They generate movement, but the movement is defensive rather than generative. Inspiration works differently. It pulls rather than pushes. It activates the brain's reward and exploration systems rather than its threat systems.

When people feel pulled toward something meaningful, curiosity rises, energy increases and effort becomes voluntary rather than extracted. The work that emerges from this state is qualitatively different from the work that emerges from pressure. It is more creative, more sustained, more original. And the difference is not motivational. It is neurological.

This is the difference between people doing the job and people wanting to do the job. That difference defines performance at every level, and it cannot be faked. The leader who relies on push will eventually exhaust their team. The leader who can generate pull will find that the team brings energy back to them.

There is a reason storytelling has always been central to leadership. The brain does not respond to instructions the way it responds to meaning. When a leader communicates a clear and compelling vision, multiple brain regions activate simultaneously. Attention increases. Emotion engages. Memory strengthens. People begin to internally simulate the future being described. They are not just hearing the words. They are experiencing a possibility. A leader who can transform technical information into a story about where the organisation is going produces a different effect from one who simply delivers updates. The first creates direction. The second creates information overload.

The other thing inspiration requires is the willingness to introduce productive discomfort. The brain prefers what is known, predictable and safe. It conserves energy by defaulting to familiar patterns. But growth does not happen in comfort. It happens in expansion, and expansion requires the introduction of something the brain has not yet integrated. Neuroplasticity, the capacity for new neural connections to form, depends on novelty and challenge. Without them, leadership becomes maintenance. The best leaders understand this without making it brutal.

They do not create chaos, but they do create constructive tension. They ask better questions than the team is used to. They challenge assumptions that have become invisible. They invite different thinking, not for the sake of disruption, but to expand what is possible. When they do this well, the result is not stress. It is sharper thinking and a sense of being trusted with more than the obvious.

Emotional contrast becomes important here. A leader who is always intense becomes exhausting. A leader who is always calm becomes unreadable. But a leader who can shift, who can be calm under pressure and urgent when urgency is needed, who can celebrate genuine success and acknowledge genuine difficulty, becomes credible. The brain pays attention to contrast. It registers shifts. In those shifts, it finds meaning. People around such a leader feel that the moment is being met accurately, and they relax into the sense that someone is paying attention.

The greatest barrier to all of this is rarely external. It is internal. Many leaders carry a narrative beneath their behaviour: I am not ready, I need to prove myself, I cannot show weakness. These narratives create tension, and tension blocks inspiration. You cannot expand others if you are internally contracted. You cannot transmit aliveness if you are spending your energy holding yourself together. The work of inspiration begins, then, with the inner work of integration. Before a leader can be a source of energy for others, they have to have access to their own.

Inspiration cannot be reduced to a technique. Communication training, vision frameworks and storytelling templates can refine expression, but they cannot manufacture the underlying state. People can tell the difference, even when they cannot articulate it. They know when they are being moved and when they are being managed.

When Satya Nadella became CEO of Microsoft, the company had immense capability but little energy. It was internally competitive, fragmented and cautious. His shift was not primarily strategic. It was cultural. He moved the organisation from certainty to curiosity, from competition to collaboration, from control to learning. He spoke openly about empathy and asked people to learn rather than to prove. This was not soft leadership. It was a recalibration of the conditions under which people thought and contributed. The measurable results, renewed innovation, stronger engagement and significant growth, followed from a change in state, not just a change in direction. That is inspiration at scale. It is not the result of charisma. It is the result of someone deciding what kind of field to create around themselves and around the work, and then living inside that decision consistently enough that others felt safe to step into it.

In the world we are now entering, this matters more than ever. As external intelligence becomes more capable, what machines cannot do is generate meaning that lands in another person's nervous system. They cannot transmit a state. They cannot make someone feel that their effort matters. The future of work will not be won by efficiency. It will be won by the quality of human energy that can be generated and sustained inside organisations. That energy does not appear by itself. It is shaped by leaders who understand that their state is not private, that their presence is not neutral, and that what they transmit becomes the condition under which others either come alive or shut down.

You cannot force imagination. You cannot command creativity. You can only create the conditions in which they emerge. And those conditions begin with a leader who is inspired enough to be inspiring. Inspiration is not something you perform. It is something you become. It is the alignment between what you believe, how you feel and how you show up. When that alignment is present, the people around you sense it before you say a word.

They listen differently. They think differently. They act differently. Not because they have been told to, but because something in them has been activated by something in you. The question is not whether you have authority. The question is whether anything in you is alive enough to move anyone else.

Reflection Question

Think about the last time you walked into a room and felt the energy shift, for better or worse, the moment you arrived.

What state were you carrying? What would the same room have felt like if you had arrived in a different state?

CHAPTER 12

IMAGINATION

THE CAPACITY TO SEE WHAT IS NOT YET THERE

Every era has its particular blindness. Ours may be the assumption that thinking faster is the same as thinking better. Across organisations and industries, the dominant mode of cognition has become procedural. Problems are solved using approaches that solved similar problems before. Decisions are made by reaching for the patterns that worked last time. Strategy meetings produce variations on what already exists rather than departures from it. The work moves forward, but it moves along grooves already worn into the ground. From the outside, this looks like competence. From a longer view, it looks like repetition. The problem is not that ideas are scarce. The problem is that the conditions for new ones have closed.

Imagination is one of the most misunderstood capacities in leadership. It is often treated as decorative, something that belongs to artists, designers and the occasional visionary founder, but not to the daily work of leading through complexity. It is associated with workshops, creative exercises and language that makes serious people slightly uncomfortable. As a result, it has been progressively demoted in most organisations, treated as a nice addition rather than a fundamental requirement. This is a mistake, and it is becoming a more expensive one.

Imagination, in the sense we mean it here, is not the production of clever ideas. It is the capacity to think in possibilities rather than only in procedures. It is the ability to move beyond what is already known and to construct, in the mind, configurations of the future that do not yet exist. It allows a leader to see beyond the immediate problem, question the framing itself and consider that the most useful answer may not be one that has been used before.

Without imagination, leadership becomes maintenance. With it, leadership becomes generative.

The neuroscience of imagination has advanced considerably in recent years, and what it reveals is striking. Imagination is not a single function located in a single part of the brain. It is a coordinated process drawing on memory, perception, simulation and self-awareness, and it depends on the brain's ability to hold these elements together while reassembling them into something new. The hippocampus, traditionally associated with memory, plays a critical role in constructing imagined scenes and future scenarios, providing the anchor in lived experience from which novel possibilities can be built (Hassabis & Maguire, 2007; Buckner & Carroll, 2007).

The default mode network, once thought of as the brain's idling state, is now understood to be one of the central engines of creative thought, supporting the kind of internal exploration that produces genuine novelty (Beaty et al., 2018). Imagination is not separate from cognition. It is one of cognition's most demanding forms, requiring the brain to hold what is, what was and what could be all at once, and to allow them to recombine into something that has not yet existed.

This is why imagination is so easily lost. As we saw in the chapter on integration, the brain under pressure narrows. It conserves energy, reaches for familiar patterns and treats uncertainty as threat. What this means for imagination is direct: a mind operating in survival mode cannot generate novelty. The capacity has not been destroyed. It has simply been crowded out by the conditions that demand immediate response.

This is one of the most consistent observations from years of working with the i4 Neuroleader Model. People often describe themselves as open, curious and innovative. They see themselves as thinkers who welcome new ideas and challenge convention.

Yet when their actual patterns are examined, a much narrower picture often emerges. Their thinking has become procedural without their noticing. Creativity has been replaced by efficiency. They have been so busy executing that they have stopped imagining, and they have not had the spaciousness to notice the difference. The capacity has not been lost. It has simply not been accessed, and over time what is not accessed becomes harder to reach.

Part of what makes this difficult is biological. Imagination requires uncertainty, and uncertainty activates the brain's threat systems. The unknown cannot be predicted, and what cannot be predicted is treated as risk. The brain prefers certainty even when that certainty leads to repetition. Part of it is also cultural. Many professional environments reward correctness over curiosity. They train people to be precise, efficient and defensible, and over time this creates an internal rule: do not say what cannot be justified, do not propose what cannot be proven, do not imagine what cannot be explained. Imagination becomes risky, not because it is wrong, but because it cannot be immediately validated. So people stop, not consciously but gradually. They begin to think only in ways that can be defended, and the space for imagination closes.

Imagination requires a different relationship with uncertainty. Most leadership environments treat uncertainty as something to eliminate quickly. But imagination depends on the willingness to hold uncertainty open long enough for something genuinely new to emerge. Premature closure is the enemy of imagination. The leader who needs to resolve every question immediately is the leader who will keep producing variations on what already exists. Real imagination is not a technique. It is a condition. It requires a mind that is not constantly under pressure to conclude.

Some years ago, we worked with a group of lawyers from a large firm. They were intelligent, accomplished and deeply trained in precision.

Their world was built on precedent, clarity and defensible reasoning. Ambiguity was something to minimise. We asked them to do something that, from their perspective, bordered on absurd. After a paired conversation about who they were and what mattered to them, each participant was asked to draw the other. Not a portrait. An impression. Something that captured the essence of what they had received.

The resistance was immediate. Several protested that they could not draw. Others tried to delay starting. But once they began, something shifted. The room went quiet, but not in the focused silence of legal work. It was the silence of attention. Heads lowered, hands moving, time slowing down. When they presented the drawings to each other, the atmosphere had changed completely. People spoke more carefully, more personally. And when each lawyer received the drawing made of them, the response was almost always the same: surprise, then recognition, then something close to relief. Several said they had not been seen that clearly in years. One said, almost reluctantly, that he had forgotten he had an inner life worth drawing.

The exercise was not about art. It was about access. It revealed that imagination had not disappeared in these highly trained professionals. It had simply been suppressed by years of conditioning into precision. And once permission was given, it returned almost immediately. This is what makes imagination both fragile and recoverable. It can be trained out of people slowly, through years of professional reinforcement, and restored surprisingly quickly when the conditions allow.

It helps here to make a distinction that the wider conversation about leadership rarely makes clearly enough. Imagination is not the same as creativity. Imagination is the cognitive ability to visualise possibilities beyond the constraints of the immediate. Creativity is the transformative process that turns those possibilities into reality.

Every brain is inherently capable of the first. The second is what happens when imagination is given the conditions, the time and the discipline to take form.

Relmi, who has spent years working at the intersection of design, neuroscience and creative practice, draws this distinction sharply, and her experience bears it out. During a seven-day neurofeedback intensive in Sedona, she observed her own brain shifting into the alpha state that researchers associate with the emergence of creative insight, the same default mode network activity this chapter has already described. What she found was not new in the literature. It was new in her body. The widespread belief that some people are creative and others are not is, in her view, a fallacy. Imagination is not a gift withheld from most and granted to a few. It is a capacity available to everyone whose system is given the conditions to access it.

There is also another dimension of imagination that leadership literature rarely names. Imagination is not only about generating new ideas. It is about being able to imagine other people. The leader who can genuinely imagine what a colleague is experiencing, what a customer is feeling or what a team member is not saying, has access to a form of intelligence that purely analytical leaders do not. This capacity, often described as self-projection, allows the brain to step outside its own immediate experience and simulate the experience of another (Buckner & Carroll, 2007).

It is one of the foundations of empathy, ethical reasoning and the capacity to lead in ways that account for more than the leader's own perspective. A leader without imagination is not only less innovative. They are less able to see other people clearly. Their decisions become technically correct but relationally blind, their strategies coherent on paper but failing in practice because they do not account for the people who must live inside them.

The contrast with artificial intelligence sharpens this. Machines remix. Imagination creates. This distinction is not theoretical. It reflects a generation that has grown up inside constant exposure to images, ideas and immediate outputs. The risk is not that creativity disappears, but that the slower, more uncertain process from which original thought emerges is no longer required. When generated material is always available, imagination is not destroyed. It is bypassed. And what is bypassed often grows weaker through neglect. Protecting imagination now means more than encouraging creativity. It means preserving the inner conditions in which something genuinely one's own can still arise.

The difference may not always be visible in the output, but it is fundamental in the process. Generated content is built from patterns that already exist in the training data. It cannot move beyond what has been encoded. Human imagination, at its deepest, is not bound by what has been encoded. It can leap. It can hold contradiction. It can construct from longing, moral conviction and the felt sense of what should be true even when no precedent exists for it. As external generation becomes faster and cheaper, originality becomes scarcer. The leaders who will matter in the next decade are not the ones who use AI most efficiently. They are the ones who can imagine what AI cannot.

This places a responsibility on leadership that is often underestimated. If imagination is being eroded by pressure, speed and the constant demand for output, then protecting the conditions for imagination becomes part of the role. Not through workshops, but through design. It means creating space for thinking that is not immediately tied to deliverables. It means tolerating the discomfort of not knowing the answer yet. It means modelling attention that allows exploration rather than only execution.

It also means taking play seriously. This sounds trivial, and it is precisely because it sounds trivial that it is undervalued. Play is not the opposite of work. It is one of the conditions under which the brain becomes able to make new connections. Stuart Brown's research has shown that the absence of play is associated with measurable reductions in creative and adaptive capacity, while its presence supports the kind of flexible, exploratory thinking from which imagination arises (Brown, 2009).

The clearest large-scale demonstration of this principle is found in the Finnish education system. Finnish children begin formal schooling at age seven, later than children in almost any other developed country. Before that, early education emphasises play, outdoor exploration and social development rather than academic instruction. Even once school begins, the days are shorter than international norms, recess is protected after every lesson, homework is minimal and standardised testing is rare.

The conventional assumption would predict weaker academic outcomes. The opposite is true. Finnish students consistently rank among the highest performers in the world on international assessments, while reporting lower stress and higher wellbeing than their peers in more pressured systems (Sahlberg, 2021). This is not because Finnish children work less. It is because the conditions under which their minds develop allow them to think more deeply when they do work. Play is not a reward for completed effort. It is the soil from which serious thinking grows.

The implication for leadership is direct. Adults who have lost the capacity for play have often, without noticing it, lost the capacity for imagination as well. Organisations that treat play as wasted time tend to produce leaders who cannot generate novelty. Cultures that crowd out unstructured thinking in the name of efficiency often discover, years later, that they have produced extraordinary executors and almost no original thinkers.

The Finnish example is a rebuke to this pattern, and it deserves to be taken seriously by anyone responsible for the conditions under which other people are expected to think.

There is one final dimension of imagination that cannot be left unsaid. Imagination is closely related to hope. Not hope as wishful thinking, but hope as the active capacity to envision a future worth working toward. A leader who cannot imagine cannot hope, and a leader who cannot hope cannot lead others through difficulty. When the present is hard, people do not move because of data. They move because of possibility. The ability to picture something better than what currently exists, and to make that picture real enough that others can see it too, is one of the deepest forms of leadership there is.

Within the i4 Neuroleader Model, imagination is positioned as a core human ability not because it is pleasant or aspirational, but because it is foundational to the capacity to innovate, adapt and remain creative under pressure. The work is not to add imagination to leadership as a supplementary skill. The work is to stop suppressing it.

Imagination is not the opposite of seriousness. It is what gives seriousness somewhere to go. Without it, even the most disciplined leadership becomes a slow kind of repetition. With it, leadership opens onto something larger than the present moment. The leader who can imagine is not escaping reality. They are expanding it. The question is not whether you have ideas. The question is whether you still allow yourself to see beyond what already exists.

Reflection Question

What is one possibility you have stopped allowing yourself to consider, because it would be hard to justify if you said it out loud?

And what would it cost you, in five years, to keep not considering it?

CHAPTER 13

INTUITION

THE INTELLIGENCE THAT ARRIVES BEFORE EXPLANATION

There is a moment every leader recognises, even if they rarely name it. A decision is required, the information is incomplete, and the analysis could continue indefinitely. Yet beneath the surface of competing data and rational arguments, something has already begun to form. A sense. A direction. An internal signal that precedes explanation. It does not arrive as logic. It arrives as knowing. Most people hesitate in that moment, not because the signal is weak, but because it cannot yet be justified. So they override it, delay it or attempt to translate it into something that feels more acceptable. This is where intuition lives. Not in the absence of intelligence, but in the space before it becomes fully articulated.

Modern leadership has been built on the strength of analytical thinking. Data, measurement and structured reasoning have transformed how organisations operate and how decisions are made. This shift has brought clarity, accountability and precision. But it has also created a subtle distortion. It has led many to assume that only what can be measured can be trusted, and that only what can be explained can be acted upon. In stable environments, this works. In complex ones, it begins to reach its limits. Because not everything that matters is immediately visible. Not everything real is immediately measurable. And not every decision can wait for complete information. This is where intuition becomes not optional, but necessary.

Intuition is often misunderstood as something vague, emotional or unreliable. Yet research across neuroscience and decision science offers a far more precise account.

Intuition can be understood as a rapid, integrative process in which the brain draws on accumulated experience, pattern recognition, emotional signals and contextual awareness before conscious reasoning has time to fully engage. What we later describe as a gut feeling is not an absence of thinking. It is thinking that has already occurred, but has not yet reached language (Klein, 2004).

It is important not to confuse intuition with imagination. Imagination is generative. It allows us to construct possibilities that do not yet exist. Intuition is perceptive. It allows us to sense patterns, risks and opportunities that are already present but not yet fully visible. One expands what could be. The other senses what is. In effective leadership, both are essential, but they serve different roles. Imagination opens the field. Intuition reads it.

Gary Klein's work with firefighters and emergency responders illustrates this clearly. In high-stakes situations, experienced professionals do not evaluate options sequentially. They recognise patterns that match prior experience and act accordingly. Their intuition is not guesswork. It is compressed expertise, operating beneath conscious awareness (Klein, 2004). This form of knowing is not confined to extreme environments. It is present in everyday leadership. Over time, leaders build internal maps of people, dynamics, risks and opportunities. These maps are rarely explicit, yet they continuously shape perception. When something feels off in a meeting, when a strategic move feels premature, or when a direction feels unexpectedly right, these responses often emerge from that accumulated pattern recognition interacting with the present moment.

What has become increasingly clear, however, is that intuition is not only a function of the brain. It is an embodied process. Research into the gut-brain axis shows that the body plays a central role in how we perceive and decide.

The enteric nervous system, often referred to as the second brain, contains a vast network of neurons and communicates continuously with the central nervous system. This bidirectional communication influences mood, perception and judgement (Mayer, 2016; Cryan & Dinan, 2012). Antonio Damasio's work reinforces this through the somatic marker hypothesis, demonstrating that bodily signals are not peripheral to decision-making but integral to it. Emotional and physiological states guide judgement long before reasoning becomes explicit (Damasio, 1994).

This reframes intuition entirely. It is not simply something we have in our head. It is something we experience through the body. A tightening in the stomach. A sense of contraction. A feeling of openness or ease. These are not metaphors. They are physiological signals generated by a system continuously integrating internal and external information.

David Leong's work on entrepreneurial intuition extends this understanding into a systems model. He describes intuition as emerging from the interaction of three centres, the head, the heart and the gut, where cognitive processing, emotional coherence and visceral sensing operate together. In this model, intuition is not irrational. It is the result of coordinated signalling across biological systems, mediated in part through the vagus nerve, which carries a significant proportion of information from the body to the brain (Leong, 2025).

From this perspective, intuition is not something mysterious that appears from nowhere. It is the expression of a system functioning in synchrony. When cognitive, emotional and physiological processes are aligned, perception becomes clearer. When they are fragmented, perception becomes less reliable. The greatest barrier to intuition, however, is not biology. It is conditioning.

Most leaders are not trained to trust what they sense. They are trained to justify what they can prove. From early in their careers, they learn that credibility depends on evidence, that decisions must be defended, and that confidence must be anchored in data. This is not inherently problematic. It becomes problematic when it is exclusive. Organisations reinforce this pattern. Leaders are rewarded for explanation, for showing their reasoning, for making their thinking visible. Over time, this creates an unintended consequence. Forms of knowing that do not arrive with immediate justification begin to feel less legitimate.

We have seen this pattern in nearly every leader we have worked with. The moment of sensing arrives. It is overridden almost immediately. The leader reaches for an explanation that comes too late, often after the situation has already shifted in the direction the original signal was pointing. By the time the analysis catches up, the opportunity to act on the perception has passed.

This is where intuition starts to erode. Not because it disappears, but because it is overridden. Leaders begin to hesitate when they sense something they cannot yet explain. They delay decisions that feel internally clear. They search for external confirmation before acting on internal coherence. Gradually, they become more dependent on validation and less connected to perception. This is not a capability problem. It is a trust problem.

Re-engaging with intuition, therefore, is not about acquiring something new. It is about restoring a relationship that has been deprioritised. This connects directly to the broader argument of this book. When the system is under continuous pressure, attention fragments and perception narrows. The issue is not that information is missing. It is that the ability to integrate it is compromised. Intuition does not compete well with noise. It requires a degree of internal stability.

This is why clarity often emerges when pressure subsides. During a walk. In moments of stillness. In natural environments. Research shows that exposure to nature reduces stress, lowers cortisol levels and enhances cognitive functioning and problem-solving capacity (Kuo, 2015; Park et al., 2010). These physiological shifts create the conditions in which subtle signals can be perceived more accurately. Many people recognise this directly. Solutions that felt inaccessible under pressure appear when the system settles, not because new information has been added, but because existing information can now be integrated. Intuition, in this sense, is less something we generate than something we allow.

There is also a relational dimension to intuition that is easily overlooked. Some of the clearest intuitive signals do not arise in isolation, but in the subtle field between people. A pause that lingers too long. A shift in tone. Something unsaid that nevertheless becomes palpable. These are not abstract impressions. They are forms of perception sharpened by attention. In this way, conversation is not only a vehicle for connection. It is also a place where intuition becomes legible. The more present we are with another person, the more finely we can sense what is emerging beneath the words.

Intuition also deepens through experience. Over time, repeated exposure to complex situations strengthens the brain's ability to recognise patterns and anticipate outcomes. The body becomes part of this learning loop, encoding signals that inform future decisions. What feels like instinct is often the accumulated intelligence of prior experience, integrated across cognitive and physiological systems. Leaders who cultivate attention, regulate their internal state and engage in reflection tend to develop more reliable intuition. Not because they rely less on thinking, but because they are able to integrate thinking, sensing and experience more effectively.

Embodiment is central to this. The ability to notice physical responses, tension, ease, contraction, expansion, reconnects cognition with the body's signalling systems. Practices that support awareness and regulation, including pauses in decision-making, mindfulness and time in restorative environments, increase access to these signals. They do not create intuition. They make it available.

These elements reinforce one another. As attention stabilises, awareness becomes more precise. Signals that were previously lost in noise begin to stand out and, with repeated recognition, trust in them grows. Over time, action becomes less dependent on external confirmation and more aligned with what is internally coherent.

At the organisational level, the implications are significant. If leaders are rewarded only for what they can explain, they will prioritise explanation over perception. If they are required to justify every decision in linear terms, they may gradually lose access to forms of knowing that do not operate linearly. This creates environments where judgement is constrained, even when capability is high. Reintroducing intuition into leadership is not about reducing rigour. It is about expanding the range of intelligence that is recognised and utilised.

This becomes increasingly relevant as technological capability expands. Artificial intelligence can process data, identify correlations and generate outputs at scale. What it does not replicate is embodied perception, the integration of lived experience, physiological signals and contextual awareness in real time. Intuition operates precisely in this space. It does not compete with machine intelligence. It complements it. As external systems become more powerful, the human capacity to read what cannot be quantified becomes more valuable, not less.

Within the i4 Neuroleader Model, intuition is one of the four core human abilities and supports agility. It enables leaders to sense emerging patterns, anticipate shifts and act with responsiveness rather than delay. Without intuition, agility becomes reactive. With it, agility becomes perceptive.

The risk is not that intuition is absent. The risk is that it is no longer recognised as a form of intelligence. Leaders sense, but hesitate. They know, but wait. They defer to what can be proven, even when something more immediate is already available. Re-engaging with intuition is not about abandoning rationality. It is about restoring balance. It is about recognising that perception often precedes explanation, and that effective leadership requires access to both. In conditions of uncertainty, the ability to sense what is emerging becomes as important as the ability to analyse what is known. This is where intuition operates.

Intuition rarely announces itself loudly. It does not argue or persuade. It appears often before the mind has caught up, offering a direction that may feel uncertain yet carries a distinct sense of coherence. The question is not whether it is present. The question is whether it is recognised.

Reflection Question

Recall a moment when you sensed the right direction before you could fully explain it. Did you trust that signal or override it?

What would change in your leadership if you began to treat that knowing as a form of intelligence rather than something to justify?

PART IV

LEADING IN AN AGE OF ARTIFICIAL INTELLIGENCE

THE QUESTION IS NO LONGER WHAT TECHNOLOGY CAN DO, BUT WHO WE CHOOSE TO BE IN RESPONSE TO IT

CHAPTER 14

COGNITIVE SOVEREIGNTY

WHO IS DOING THE THINKING?

The changes we are living through do not arrive gently. They do not wait for institutions to prepare, for cultures to adapt or for individuals to feel ready. They move faster than our habits of understanding, and faster than the language we have relied on to make sense of leadership itself.

Between us, we carry more than fifty years of teaching, coaching and writing about leadership. Over those decades, we have watched the field evolve through countless models, frameworks and interpretations. Competencies have been mapped. Behaviours codified. Entire systems designed to help people think more clearly, relate more effectively and act with greater awareness. Organisations have invested in this work seriously, believing that better leadership could translate into better outcomes, better cultures and, ultimately, a better world. And yet, if we step back and look honestly, something does not hold. If leadership has been studied so thoroughly, refined so consistently and taught so widely, why does it still feel so unevenly lived? Why does the quality of leadership we encounter across institutions and societies still fall short of what this moment asks of it?

The question is not rhetorical. The knowledge exists. The tools exist. The language exists. But when those tools meet reality, something thins. Something fails to reach the place where it would actually change how a person thinks, speaks or decides. Which opens a more demanding possibility: perhaps the shortfall is not in leadership development at all. Perhaps the conditions in which leadership now operates have shifted more radically than we have acknowledged.

We are no longer leading in the same environment in which most leadership ideas were formed. We have crossed into a phase where intelligence itself is no longer exclusively human.

That sentence is easy to read and easy to underestimate. But it marks a structural change in what it means to be a thinking creature. For most of human history, intelligence was part of the boundary that defined us. The ability to analyse, interpret, generate language and make sense of complexity was not simply useful. It was central to how we recognised ourselves. Now that boundary has moved. Systems outside the human mind can perform many of these functions with increasing fluency and speed. The shift is not only technological. It is experiential. When intelligence becomes available on demand, the relationship we have with our own thinking begins to change. The effort required to arrive at understanding decreases. The time between question and answer collapses. And gradually, almost imperceptibly, the space in which thinking develops begins to narrow.

This is where the question of leadership turns. For decades, leadership development has focused on capability: how to think more clearly, decide more wisely, communicate more effectively and perform under greater pressure. Those questions remain valid, but they no longer reach far enough. When thinking itself can be supported, accelerated or partially externalised, capability alone cannot define leadership. A more fundamental question emerges. Who is doing the thinking? And at what point does assistance become substitution?

This is where cognitive sovereignty comes into view. The term may sound unfamiliar, but the experience it describes is not. It refers to the capacity to remain the author of one's own mind, to reason without being subtly directed, to decide without surrendering judgement and to engage with powerful tools without dissolving into them.

Work at the intersection of neuroscience, ethics and artificial intelligence has begun to treat this not as an abstract ideal, but as something requiring active protection. Ienca and Andorno (2017) have argued that mental integrity and cognitive liberty must now be considered alongside more established human rights, precisely because emerging technologies have the capacity to influence not only behaviour, but the processes that generate it. The shift is subtle, and it changes everything. The frontier of autonomy is moving inward.

This is not only a legal concern. It shows up in the texture of everyday life, in the way attention is guided often without awareness, in the way options are framed before we have fully considered them and in the way conclusions appear ready-formed, inviting acceptance rather than engagement. Zuboff (2019) made clear that digital systems do not simply respond to human behaviour. They anticipate it. They learn from it. Increasingly, they are designed to influence it. Behaviour is no longer only observed. It is shaped.

Han (2017) describes a related movement from another angle. Control no longer operates primarily through restriction. It operates through alignment, through the shaping of preference, the steering of attention and the subtle modulation of how individuals experience themselves. Influence becomes internal. It does not feel like force. It feels like ease.

This is why the shift is so difficult to detect. Nothing appears to be taken away. Friction disappears. Speed increases. Answers arrive effortlessly. At no point does it feel as though something has been lost, which is precisely how the loss occurs. The danger is not that machines become more intelligent. The danger is that human beings become less deliberate. Not less capable, but less present in the act of thinking itself. A person can use powerful systems and remain fully engaged in their thinking. But they can also begin, gradually and without noticing, to hand over more than effort.

They can hand over reflection. They can hand over interpretation. They can begin to accept what is presented before entering the process themselves. Over time, thinking becomes lighter. Faster, but thinner. And leadership cannot function well on thin thinking.

Leadership requires something different. It requires the ability to stay with complexity without collapsing it prematurely, to recognise when something appears correct but does not feel complete, to notice what is absent and not only what is present, and to remain with a question long enough for it to deepen rather than closing it at the first available answer. These are not efficiencies. They are acts of presence. This is where cognitive sovereignty becomes lived rather than conceptual. It is not about rejecting technology. It is about remaining in relationship with one's own thinking while using it. It is the difference between being supported and being replaced.

There is a particular moment that reveals whether this capacity is intact, and it rarely looks dramatic. A decision needs to be made. Information is available, or easily accessible. A system can generate options, summarise possibilities, even suggest a direction. The process moves quickly. The path forward appears almost immediately. And yet something subtle is at stake. Does the leader pause long enough to think, or move directly to output? Do they examine what is offered, or accept it because it appears plausible? Do they stay with the question, or allow it to be closed on their behalf? These moments are small, but they accumulate, and over time they form a pattern. Cognitive sovereignty is not lost in a single decision. It erodes through repetition.

We write this not as distant observers, but as people living inside the same conditions we are describing. Having worked closely with artificial intelligence over the past two years, we have felt, directly and often, how easily one can be drawn into the pace algorithms offer.

When the working day ends, it becomes strangely difficult to walk down to the kitchen and wash the dishes at the pace of ordinary life. Walking without a phone feels unfamiliar. Going to sleep without checking one more message asks for more discipline than it used to. These are small things, almost too small to mention, and that is precisely what makes them significant. The pull is not dramatic. It is constant. It shapes the rhythm of attention long before it shapes the content of thought.

What becomes clear in those moments is that the discipline of our time is not primarily about managing information, tools or output. It is about managing our minds, our cognition and our brains more deliberately than ever before. The question is no longer whether the technology is useful. It clearly is. The question is whether we can remain internally organised while using it, whether we can return to slower rhythms when the day ends, whether we can still tolerate the spaces in which thinking deepens and the self recovers. This is not a nostalgic concern. It is a practical one. Because the capacities leadership depends on, attention, patience, discernment, presence, are shaped by what we do in the hours when no one is watching, including the hours we spend with our devices.

This is why awareness alone is not enough. Most people would say they value independent thinking. They would say they do not want to be influenced without their consent. But the lived conditions of modern work tell a different story. Speed is rewarded. Completion is prioritised. The ability to move quickly becomes more visible than the ability to reflect deeply. What becomes optional eventually becomes rare. The difficulty is not that people no longer think. It is that they are given fewer reasons to think deeply.

In earlier environments, understanding had to be earned. Information was sought, held, compared and integrated. That process was slower, but it strengthened the very capacities leadership depends on: attention, patience, discernment and the willingness to remain with uncertainty. Now much of that process can be bypassed. Answers appear before questions have fully unfolded. The cost of access has decreased, but so has the necessity of engagement. The mind shifts from a place where thinking is generated to a place where outputs are received.

For a generation that has grown up with immediate access to answers, this shift becomes even more pronounced. The question is no longer only what we know, but whether we still engage deeply enough to think. As Relmi has often reflected in her work, the risk is not that knowledge disappears, but that thinking becomes optional. And when thinking becomes optional, connection begins to thin. It is through thinking together, not simply accessing information, that we remain human. At first, this feels like progress. It removes friction and reduces effort. But over time, it can reduce depth, not because depth is no longer possible, but because it is no longer demanded. And leadership, at its core, depends on depth.

This is where the connection to the previous part of the book becomes essential. Integration allows a leader to remain steady enough to notice what is happening. Inspiration allows them to generate energy rather than depend on constant stimulation. Imagination allows them to move beyond what is presented. Intuition allows them to sense what has not yet been made explicit. Without these capacities, cognitive sovereignty becomes fragile. A fragmented system is easier to influence. A depleted mind defaults more quickly. A procedural thinker follows what is given. A disconnected leader operates from borrowed ground. This is not a failure of intelligence. It is a loss of authorship.

The consequences extend beyond the individual. A leader who does not think deeply cannot create a culture that thinks deeply. A leader who defers judgement cannot expect others to exercise it. A leader who prioritises speed over reflection will shape teams that optimise for completion rather than understanding. From the outside, such systems may still appear efficient, even high performing. But beneath that surface, something is absent. People respond quickly but question less. They produce consistently but originate less. Over time, this narrows the organisation's capacity to adapt, to innovate and to meet complexity in ways that are not already encoded.

This is why cognitive sovereignty is not optional. It is not a philosophical preference. It is a condition for responsible leadership in environments where influence operates at the level of thought. And it returns us to the question sitting beneath this entire part of the book. Not what technology can do, but who we are becoming in response to it.

If leadership is to remain meaningful, it will not be defined only by what a person can produce with increasingly powerful systems. It will be defined by whether they can remain internally present, ethically grounded and cognitively self-directed while using them. The future of leadership will depend less on how much intelligence we can access, and more on how much humanity we can sustain while accessing it. That is a higher standard, but it is also the one this moment requires. The external world is changing rapidly. The internal world must not be left unattended. And if leadership is to remain worthy of the systems we are creating, it will have to begin there.

Reflection Question

Think of the last time an answer arrived before you had time to sit with the question. A summary, a suggestion, a direction that appeared almost ready-made.

Did you stay with it long enough to notice whether it was actually yours? Or did you accept it because it sounded right and move on?

CHAPTER 15

PURPOSE AFTER AUTOMATION

WHEN WORK NO LONGER GUARANTEES IDENTITY

If the previous chapter asked who is doing the thinking, this one asks a harder question. What happens to people when the thinking, and the doing, are no longer entirely theirs?

He had been in the role for twenty-three years. The systems changed slowly at first. Small efficiencies. Better tools. Tasks that once required hours of concentration could now be completed in minutes. At the beginning, it felt like progress. Less friction. More output. A clearer path through the day. Then something shifted.

The work did not disappear all at once. It thinned. Decisions that once required judgement were now suggested automatically. Reports assembled themselves. Patterns surfaced before he had time to look for them. What remained for him to do became narrower, more supervisory, less essential. He still had a title. He still had responsibilities. From the outside, very little had changed. But something inside the experience had. The sense of being needed began to loosen. The effort that once structured his days no longer carried the same weight. What had previously required attention, skill and presence now required oversight. He was still working, but the work no longer seemed to need him in the same way. And with that, a question began to surface. If the work continues without me, what exactly am I here for?

The question is no longer abstract. It is surfacing across professions, industries and levels of seniority, and it does not always arrive clearly. Sometimes it appears as restlessness. Sometimes as disengagement. Sometimes as a vague sense that something once meaningful has become mechanical. What is shifting is not only the structure of work. It is the relationship between work and identity. For a long time, that relationship felt stable. Work provided income, certainly, but also something less visible and more powerful. It offered structure, recognition and a way of understanding one's place in the world. To work was not only to produce. It was to belong, to contribute, to matter.

Susskind (2020) has argued that work has historically served multiple hidden functions beyond wages. It has distributed not only income, but also purpose, direction and social identity. When work changes, those functions do not automatically relocate. They destabilise. What appears as an economic shift quickly becomes a psychological one.

Automation does not simply change how work is done. It changes what it means to be the one doing it. As more tasks become automated, the effort required to produce outcomes decreases, but with that, something else begins to loosen. When effort is no longer tightly coupled with value, the old equation between work and worth begins to break.

This does not happen dramatically. It happens gradually. The role remains. The outputs continue. But the internal experience shifts. What once required presence becomes procedural. What once demanded engagement becomes supervision. What once carried meaning becomes efficient. And efficiency, on its own, does not generate purpose.

This is where the traditional idea of purpose begins to fracture. Modern culture has equated purpose with productivity. To be purposeful has meant to be useful, to contribute in measurable ways, and to generate outcomes that can be recognised and rewarded. Under this model, work becomes the primary site of meaning. The problem is not that this model is entirely wrong. It is that it is incomplete. When purpose is tied too tightly to work, any disruption to work destabilises identity.

If the role changes, the sense of self shifts with it. If the role disappears, something more fundamental is called into question.

Danaher (2019), in his work on post-work futures, makes a crucial distinction here. He argues that human flourishing cannot remain dependent on labour alone. If work becomes less central, then meaning must be reconstituted elsewhere, through activities, relationships and forms of engagement not defined by necessity or productivity. This is not a distant scenario. It is already beginning. The question is not whether work will disappear entirely. It is whether it will continue to carry the weight of meaning it once did. Increasingly, it cannot.

Which leads to a more precise distinction. Purpose is not created by work. It is expressed through it. This changes where we look. If purpose is created by work, then the loss or transformation of work threatens meaning directly. If purpose is expressed through work, then work becomes one of several possible channels through which meaning can take form. This reframing becomes essential in a world where work is no longer stable enough to carry identity on its own.

Juan Carlos has long described this shift in terms that place it inside a much larger historical arc. Drawing on his concept of talentism, he argues that we are living through a change as structural as the transition from feudalism to capitalism. In the feudal era, value was tied to land. In the capitalist era, value was tied to capital. Now, value is shifting toward talent, toward what a person can think, create and contribute that cannot be replicated by systems. But talent, in his framing, is not simply a set of skills. It is the convergence of vocation, fluidity and the willingness to keep learning. When a person has not yet discovered their vocation, the thing they are genuinely made for, no amount of productivity will generate the sense of meaning they are looking for.

And when organisations fail to recognise that talent is now more scarce and more valuable than capital, they continue to manage people as though they were interchangeable, which is precisely the assumption that automation exposes as unsustainable.

Across decades of executive coaching, he has watched leaders who operate from genuine vocation sustain purpose through every kind of disruption, while leaders who have built their identity entirely around their role find themselves hollowed out when that role begins to change. The question of purpose after automation, from this perspective, is not new. It is the latest expression of a shift that has been underway for decades, and the leaders who will navigate it best are those who have already begun to locate their value in what they are, rather than only in what they do.

Acemoglu and Johnson (2023) add an important layer. In their analysis of technological progress, they show that technology does not automatically enhance human flourishing. It can displace, deskill and narrow human roles if it is directed primarily toward efficiency rather than complementarity. What matters is not just that technology advances, but how it is integrated into human life. If systems are designed only to optimise output, the human role risks becoming increasingly peripheral. And when people feel peripheral, meaning erodes, not because they lack capability, but because their relationship to what they are doing grows thinner. The question of purpose after automation is not only about replacing jobs. It is about restoring depth.

To see this more clearly, it helps to step outside highly technologised environments. On the Indonesian island of Sumba, not far from Bali, we have seen a different relationship between work and meaning firsthand. Traditional houses stand tall against the landscape, built not for efficiency but for continuity. People gather around shared rituals, land and relationships.

Work exists, of course. It is necessary, physical and continuous. But it is not treated as the sole carrier of identity. People contribute because they are part of something, not because contribution is the only way to justify their existence. The difference is not romantic. It is structural. Purpose, in that context, is distributed. It lives in relationships, traditions, shared time and continuity with something larger than the individual. It is not concentrated in a single role. This does not mean modern societies must replicate that way of life, but it reveals something that has been obscured. Human meaning has never depended on output alone.

We are now in a position where that truth is becoming unavoidable. As automation absorbs more of what once defined work, the question is not only what tasks remain, but what remains of the person when those tasks no longer anchor identity. This is where the conversation must move. Not toward replacing lost roles as quickly as possible, but toward understanding what those roles were holding in place. Across years of working with leaders, a pattern appears. When people begin to experience this shift, the immediate response is to search for a new role, a new structure and a new way of restoring the previous equation between effort and meaning. But the deeper movement is different. It is not about replacing one role with another. It is about re-examining where meaning actually lives.

What emerges, when this is explored carefully, is that purpose tends to organise itself around three interrelated conditions. The first is physiological. A person whose system is depleted struggles to experience meaning, regardless of external success. Chronic fatigue, disrupted sleep and sustained stress narrow perception and reduce the capacity to feel engaged with anything beyond immediate demands. Meaning requires energy. Without it, even meaningful work feels empty.

The second is interpretive. Meaning is not inherent in activity. It is constructed through the lens a person brings to it. Two individuals can perform the same task and experience it entirely differently depending on how they understand its significance. When interpretation is outsourced to status, approval or expectation, meaning becomes unstable. The third is directional. Purpose requires movement, not a fixed, grand mission, but a sense of orientation toward something that matters. Without direction, activity becomes repetitive. With direction, even small actions begin to accumulate into coherence.

It is also worth remembering that brains, temperaments and longings differ profoundly from one person to another. Purpose does not take a single shape, and the transition being described here will not look the same for everyone. Some people love working and, regardless of what they do, will always want to keep themselves occupied. For them, activity itself is a source of life, and the question is less whether to work and more how to choose work that remains meaningful.

Others feel most themselves in contact with nature and find the sedentary rhythms of white-collar life corrosive, a form of life that, according to many observers of artificial intelligence, may in any case become less central over time. Others come alive through contact with people, through the daily work of helping, listening and being present to others, and would lose something if that were taken away. Others again are most at home in the act of creating, returning to the blank page, the empty studio or the unformed idea, finding meaning less in any particular outcome and more in the ongoing capacity to bring something new into being.

None of these orientations is more legitimate than another. They are expressions of different nervous systems, different histories and different internal landscapes.

What matters is not which form of purpose a person inhabits, but whether they remain connected to the one that is truly theirs. The danger of automation is not that it will eliminate meaning, but that it may pull people further from the particular shape of meaning that belongs to them, offering instead a generic version shaped by convenience, comparison or default. Recovering purpose, in this sense, is not only about finding something to do. It is about recognising what one is actually made for, and giving that recognition room to live.

These three conditions do not create purpose. They allow it to be experienced. When one of them weakens, something inside the person goes quiet. This is often misinterpreted as a lack of motivation. It is more accurately a loss of alignment. And this is where the earlier part of the book becomes essential again. Integration stabilises the system so that meaning can be felt. Inspiration reconnects the individual to internal energy rather than external pressure. Imagination allows new forms of contribution to emerge when old ones dissolve. Intuition helps navigate direction when clear structures are no longer available. Without these capacities, the shift brought by automation can feel like loss. With them, it becomes transition.

This is why the question of purpose after automation is not only about work. It is about identity. It is about whether a person's sense of value remains dependent on external structures, or whether it can be sustained across changing conditions. The human spirit is often described as something abstract, something difficult to define. But it leaves traces. In what people return to when no one is asking. In what continues to matter when recognition is removed. In how they choose to spend time when productivity is no longer the only measure.

When those traces fade, people do not immediately collapse. They compensate. They work harder. They optimise more. They search for new ways to feel useful. But usefulness is not the same as aliveness. And aliveness cannot be restored through efficiency. It returns through alignment.

The future of work will continue to evolve. Roles will change. Entire categories of activity will emerge and disappear. But beneath those shifts, a more stable question remains.

Where does your sense of meaning reside when it is no longer held in place by what you do?

That question cannot be answered by automation. It has to be lived.

Reflection Question

Picture a version of yourself ten years from now, in a world where much of the work you do today is carried out by systems.

What are you doing with your hours? Who is near you? What kind of effort are you pouring yourself into that you would still call meaningful?

And if that version of you feels a little unfamiliar, even uncomfortable, ask yourself what it would take to start recognising them, not by waiting for the world to change, but by giving that self one small place in your life today.

CHAPTER 16

RAISING HUMANS IN THE AGE OF AI

WHAT MUST NOT BE OUTSOURCED

If the previous chapter asked where meaning lives when work can no longer hold it, this one turns toward something even more unsettling. What happens to the people we love most, the ones still being formed, when the very conditions of formation begin to change?

The question is often asked too late. It arrives in moments of unease, when something feels off. Attention seems thinner than it used to. Responses are faster but less considered. Children appear capable, articulate and well informed, yet somehow less anchored in themselves. By the time the question is spoken clearly, much of the shaping has already taken place.

What are we preparing the next generation for?

For a long time, the answer felt stable. We prepared them for knowledge, for professions, for entry into systems that rewarded what they could accumulate, remember and apply. The logic was simple and widely accepted: the more a person knew, the more capable they would become. That logic is no longer sufficient. Not because knowledge has lost value, but because it has lost its exclusivity. Information is now ambient. Answers are immediate. The effort that once sat between question and understanding has been compressed and, in some cases, removed entirely. A child no longer needs to wrestle with a problem for long before meeting a solution.

At first glance, this looks like progress. But something important is shifting underneath it. Learning was never only about arriving at the answer. It was about what happened while the answer was being formed: the hesitation, the partial understanding, the frustration, the rethinking, the moment when something finally made sense because the mind had stayed with it long enough. These were not inefficiencies. They were the process through which the mind organised itself. Without that process, knowledge can still be acquired, but the architecture that supports deeper thinking may not be built in the same way. The child becomes informed, but not necessarily formed.

The challenge, then, is developmental rather than informational. We are raising human beings in a context where cognition itself is continuously supported by external systems. Tools can suggest, generate, summarise and guide. The risk is not that children will know less. It is that they will practise less of what thinking actually requires. The ability to think independently does not emerge from exposure. It emerges from engagement, and engagement is what disappears first when shortcuts are always available.

What matters is not whether young people use artificial intelligence, but how they use it. It can become a shortcut to ready-made answers, something that reduces effort before effort has had the chance to do its developmental work. Or it can become a stimulus for deeper engagement, prompting the mind to question, compare, test and think more actively. That distinction is decisive. Because if AI becomes a substitute for thinking rather than a stimulus for it, the developmental trajectory shifts. The child becomes efficient, but less generative. Capable, but less exploratory. Quick, but less deep. And depth is not automatic. It is built through friction.

This is the tension at the heart of the current moment. Much of modern technology is engineered to reduce friction. Faster responses. Smoother interactions. Fewer obstacles between intention and outcome. These are powerful advancements. But when applied to a developing human being without awareness of what friction actually builds, they remove the very resistance through which internal capacity is formed. One of the most important of those capacities is the ability to sit with not knowing. In earlier environments, uncertainty had duration. A question did not immediately collapse into an answer. The mind had time to expand around it. Possibilities were considered, rejected and reshaped. That space was not empty. It was generative. Today, it can disappear almost instantly, and with it something begins to fade.

Research on learning in rapidly changing environments makes a related point. Winthrop and Barton (2018), working through the Brookings Institution, have shown that the future will not reward those who simply acquire knowledge, but those who can adapt, learn continuously and navigate unfamiliar situations with resilience. These are not informational skills. They are cognitive and emotional capacities built through lived experience. If the experience changes, the outcome changes.

The same applies to attention. Attention is not a fixed trait. It is trained. It develops through repeated engagement with tasks that require focus, return and persistence. When environments become saturated with rapid stimuli and constant interruption, the brain adapts. It becomes more responsive, but less sustained. More alert, but less anchored. Over time, depth begins to feel like effort, not because the young person lacks ability, but because the conditions for developing sustained attention have been replaced by conditions that reward fragmentation.

Work on early childhood development reinforces this concern. Children learn most effectively through rich human interaction, through conversation, play and responsive engagement. Technology can simulate interaction, but it does not replicate the complexity of human exchange. Children require relational depth, not just informational input, for cognitive and emotional development to unfold fully. This is not an argument against technology. It is an argument for proportion. A child raised in environments dominated by mediated interaction may still learn, but the texture of that learning changes.

The subtle cues of human presence, the unpredictability of conversation and the negotiation of meaning in real time are not easily replaced. Without them, something in the development of social understanding and inner reflection becomes less robust.

There is a more personal layer to this, and it is the layer we want to speak from directly. After raising two children who are now adults, we can say with a clarity that does not come from theory that they grew up at a particular moment in time. When they were teenagers, social media was only beginning to emerge. Platforms like MySpace were just appearing. The digital world had not yet fully entered the intimate spaces of identity formation, friendship and self-worth. They stood at the edge of that shift. They were not untouched by technology, but they were not shaped by it in the way younger generations now are. Their sense of self developed in environments that still allowed for more privacy, more unmediated interaction and more room simply to be without being continuously exposed or measured against others. Looking back, it feels as though they passed through a narrowing window, because what has followed is something entirely different.

To be a parent today is not only to raise a child. It is to manage an environment that did not exist before, an environment designed by others, evolving continuously and operating with a level of influence that is difficult to see, let alone fully control. Parents are no longer only guiding behaviour. They are filtering systems, negotiating with platforms and setting boundaries around tools engineered to capture attention, shape habits and influence perception. This responsibility did not arrive gradually. It appeared suddenly, without training, without clear guidance and without the kind of societal structures that usually accompany shifts of this magnitude. What was once part of the external environment has moved into the developmental core of childhood.

And parents, already carrying the complexity of raising another human being, have been asked to take on an additional role: not only caregiver, not only educator, but regulator of an invisible architecture shaping their child's inner world. It is an extraordinary demand, and one that is often underestimated.

The challenge is not only about limiting exposure or setting rules. It is about something subtler. How to help a child form a sense of self that is not entirely dependent on external validation. How to protect the development of attention in environments designed to fragment it. How to allow identity to emerge before it is continuously mirrored, measured and influenced. These are not small questions, and they do not have simple answers. But they return us to something even more confronting.

There is another pattern that deserves to be named. Some of the most meaningful work we have done has been with younger generations, and we do not share the reflexive disappointment so many adults express when speaking about them. That response is troubling. It is easy to complain about young people. It is much harder to ask what kind of guidance they are actually receiving. Across years of building leadership development work with younger people, and through long experience in recruitment and coaching, one thing has become unmistakable: younger generations are not a problem to be criticised from a distance. They are still forming, still open and still capable of remarkable growth when they are met well.

That is why it is so unsettling to hear people speak about younger generations with dismissal or frustration. The question that immediately arises is simple. Are they coaching them? Are they teaching them? Are they guiding them? From what we have observed, in many cases, the answer is no. And that matters, because criticism without guidance is laziness disguised as wisdom. We do not get to complain about the generations we are failing to form.

If younger people appear distracted, uncertain or dependent on external validation, then the honest question is not only what technology is doing to them. It is what adults have stopped doing for them.

This is where the responsibility becomes unavoidable. We are not only raising children. We are shaping the conditions under which humanity will continue. The question is not whether the next generation will be capable. They will be. The tools available to them will expand what is possible in ways we are only beginning to understand. The real question is different. What kind of human will that capability be built upon? Will it be a human who can remain present without constant stimulation, who can think without immediate assistance, who can hold uncertainty without rushing to resolve it, who can imagine beyond what is given, who can sense what is not yet explicit? Or will it be a human whose intelligence is largely externalised, efficient, responsive and highly capable within systems, but less grounded outside them?

The answer will not be determined by technology alone. It will be determined by the environments we create and the choices we make within them. Every adult becomes, in some sense, a developmental architect through small decisions. When to intervene and when to step back. When to answer and when to let the question deepen. When to prioritise speed and when to protect depth. When to provide tools and when to allow the mind to work without them. These decisions do not feel significant in isolation, but they accumulate, and over time they shape a generation.

We often assume that preparing children for the future means giving them more. More knowledge. More access. More tools. More capability. But the future may not require more in the way we expect. It may require something more precise. Not only knowledge, but adaptability. Not only skills, but depth. Not only performance, but humanity. And humanity is not something that can be engineered externally. It has to be grown from the inside.

Reflection Question

Bring to mind one young person whose life you have the chance to touch, a child, a grandchild, a niece or nephew, a younger colleague, a student, someone you mentor. Not their generation in the abstract, but them specifically.

When was the last time you were fully present with them, without a screen in the room and without an agenda?

And if you cannot remember, ask yourself this: what would it cost you, practically, to offer them thirty minutes this week in which you were simply there, not correcting, not advising, just paying attention?

That small experiment may tell you more about your own capacity for presence than it tells you about them.

CHAPTER 17

BECOMING A BRAIN-FRIENDLY LEADER

WHAT IT MEANS TO LEAD FROM COHERENCE

If the previous chapter asked what we are preparing the next generation to become, this one asks what we are preparing ourselves to become, because the two questions are not separate. The adult who cannot remain present cannot teach a child to. The leader who cannot sit with uncertainty cannot create environments in which others learn to. Leadership, in this sense, is not only what we express outward. It is what we model for those still forming around us.

There is a point, toward the end of any serious inquiry into leadership, where the question itself changes. It is no longer about what leaders should do. It becomes about what allows a human being to function well enough to lead at all.

Across this book, that question has been approached from different angles. We have looked at the human system beneath behaviour. We have examined what happens when that system is placed under continuous pressure. We have followed the consequences of fragmentation, acceleration and externalised intelligence. And we have named four capacities that begin to restore what is lost when the system becomes misaligned: Integration, Inspiration, Imagination and Intuition. What remains now is not to add another model. It is to return to the beginning, but at a deeper level. Because leadership, when stripped of language and performance, is not an activity. It is a condition. And that condition determines everything that follows.

In the work we have developed over many years, brain-friendly leadership was never intended as a slogan or a stylistic variation on existing approaches.

It emerged from a more difficult question, one that sat beneath all behavioural models: what conditions allow a human being to think clearly, relate well and act with coherence under pressure? That question changes the entire orientation of leadership. It moves the focus away from performance alone and toward the system producing that performance. It recognises that behaviour is not the starting point of leadership, but its final expression. And it acknowledges something both simple and often ignored: people do not lead well when their internal systems are working against them.

The external world has become faster, more complex and more cognitively demanding, but the human system has not changed at the same pace. It still requires regulation, attention, recovery, meaning and relational connection to function well. When those conditions are not met, leadership does not disappear. It degrades. What appears on the surface may still look competent. Decisions are made. Communication continues. Output is produced. But something is missing. Presence becomes thinner. Thinking becomes more reactive. Relationships become more transactional. Over time, leadership shifts from something generative to something performative. And performative leadership cannot hold complexity for long.

This is where the distinction becomes critical. Traditional leadership development asks what a leader should do. Brain-friendly leadership asks what state a leader must be in for those actions to be available. That is not a philosophical shift. It is a biological one. The evidence from neuroscience and the learning sciences points in the same direction, though the voices making the case are many. Chronic stress narrows the very capacities leadership depends on. Under sustained pressure, the brain becomes more efficient at defending familiar patterns and less able to explore new ones. What looks like stubbornness or resistance to change is often a system protecting itself. Attention narrows. Cognitive flexibility decreases. Creative associations become harder to form. None of this is character. It is condition.

The same body of evidence tells us that attention itself is fragile. It cannot be endlessly interrupted without cost. When information is delivered in overwhelming or continuous streams, comprehension thins, retention suffers and the brain struggles to organise what it receives into meaningful structures. In leadership contexts, this shows up as teams that hear more but understand less, process more but integrate less. And across all of this runs an even more fundamental finding: learning, decision-making and behaviour are inseparable from emotional and physiological states. A brain that does not feel safe cannot think at its highest level. A system that is overloaded cannot sustain deep learning or creative thought. These are not preferences. They are constraints. Taken together, they lead to a single conclusion. Leadership effectiveness is not only a matter of skill. It is a function of how well the human system is supported to operate.

This is why the four capacities introduced earlier in the book are not optional enhancements. They are structural.

Integration is what allows the system to remain coherent under pressure. Without it, thinking fragments, attention narrows and response becomes reactive.

Inspiration is what generates energy from within rather than relying on constant external stimulation. Without it, motivation becomes dependent and unsustainable.

Imagination is what allows the leader to move beyond what is given, to construct possibilities rather than repeat patterns. Without it, leadership becomes maintenance.

Intuition is what allows perception to operate beyond explicit data, integrating experience, context and embodied signals. Without it, decision-making becomes slower, more uncertain and often less accurate in complex environments.

Together, these capacities do not describe an ideal leader. They describe a functioning human system. And that is where leadership must now begin.

The temptation, at this point, is to turn these ideas into practices, habits or prescriptions. To ask: what should I do differently? That question is understandable, but it returns too quickly to behaviour. A more useful question is simpler: What is the condition of the system I am leading from?

Because from that condition, everything else follows. If the system is chronically overstimulated, the leader will seek simplicity, often prematurely. If it is depleted, the leader will conserve energy, avoiding complexity rather than engaging with it. If it is fragmented, the leader will struggle to hold multiple perspectives, even when they understand their importance. If it is externally driven, the leader will depend on pressure rather than generating direction. None of this requires intention. It happens automatically.

Which is why awareness becomes the first movement of brain-friendly leadership. Not awareness as an abstract concept, but as the ability to notice, in real time, the state from which one is operating. This is where leadership becomes deeply personal, not in the sense of self-expression, but in the sense of responsibility. Because the leader's internal state does not remain internal. It shapes the environment around them. It influences how others think, how they respond, how they contribute. It becomes part of the system in which other people are trying to function. A dysregulated leader does not only experience fragmentation. They distribute it. A coherent leader does not only experience clarity. They enable it.

And in enabling it, they also guide, because leadership is not only a state. It is a form of care extended to those still forming around us, whether those others are children, younger colleagues or teams who have never been shown what coherence under pressure actually looks like. This is not metaphor.

It is transmission. The implications extend beyond the individual. Having worked with many organisations over the years, we have come to believe that what people call culture is often the accumulated state of the nervous systems within it. Organisations are not neutral structures.

They are environments that either support or undermine the functioning of the people working inside them. A brain-friendly organisation is not defined by perks, policies or surface-level wellbeing initiatives. It is defined by whether its conditions allow human systems to operate effectively. Does it protect attention, or continuously fragment it? Does it allow time for thinking, or compress all processes into urgency? Does it create psychological safety, or maintain subtle threat? Does it reward reflection, or only speed? Does it support recovery, or treat depletion as normal? These questions are not peripheral. They determine whether people can access the very capacities leadership depends on.

What is true in classrooms is true in boardrooms. Environments must align with how the brain actually functions if meaningful development is to occur. If the environment contradicts the functioning of the brain, performance may still be extracted, but at a cost. Over time, that cost accumulates in the form of burnout, disengagement, reduced creativity and declining quality of decision-making. What we call culture is often the collective result of these conditions. And leadership, at its most mature, is the ability to shape them.

This brings us to the final shift. For a long time, leadership has been associated with effort, with striving, pushing and overcoming, managing increasing complexity through increased exertion. That model is reaching its limits. Not because effort is no longer required, but because effort alone cannot compensate for a system that is not functioning well. The future of leadership will depend less on how much effort a person can sustain, and more on how much coherence they can access.

Coherence does not mean ease. It means alignment. It means the brain, body and emotional system are working together rather than against each other. It means attention can be directed rather than constantly pulled. It means energy can be generated rather than continuously depleted. It means thinking can remain open long enough to meet complexity, rather than closing prematurely under pressure. This is what allows leadership to move from reaction to presence. And presence is not passive. It is the most active state available. A leader who is fully present can perceive more, integrate more, respond more accurately and influence more effectively than one operating from fragmentation, regardless of how hard that person is trying.

We are entering a world where intelligence is increasingly externalised, where systems can generate answers, process information and perform cognitive tasks at a scale and speed no individual can match. The capacities we asked about in the previous chapters, the ability to stay with a question, to think without immediate assistance, to hold uncertainty without rushing to resolve it, are not only developmental concerns for children. They are the same capacities that determine whether a leader can function well in a world where intelligence is increasingly externalised.

What we protect in the young, we must also protect in ourselves. In that world, the value of leadership will not be defined by access to intelligence alone. It will be defined by what remains human in the presence of that intelligence. The ability to think independently while using powerful tools. The ability to remain grounded when everything accelerates. The ability to create meaning, not only output. The ability to relate, not only transact. These are not secondary qualities. They are the foundation of what leadership must now become.

Becoming a brain-friendly leader is not a destination. It is a continuous process of returning. Returning to awareness when attention drifts. Returning to coherence when the system fragments. Returning to meaning when activity becomes empty.

Returning to presence when speed takes over. It is not about becoming someone different. It is about becoming more aligned with how you function at your best, and doing so consistently enough that others can rely on that state. Because in the end, leadership is not defined by what you say. It is defined by the conditions you create, within yourself and around you.

Reflection Question

Think about the hour before your most important meeting last week. What were you doing? What were you reading, watching, consuming, responding to?

Were you preparing your nervous system to arrive in that room, or were you arriving with whatever state the previous hour had handed you?

And if the honest answer is the second, ask yourself what one small shift in the hour before would make you a different kind of presence in the room afterward, and whether you are willing to try it just once, this week, to find out.

CONCLUSION

A NEW LEADERSHIP PREMISE

Leadership does not begin where it becomes visible. It does not begin with behaviour, communication or the actions others can observe and evaluate. It begins earlier, and deeper, in the patterns that shape our internal state, in the biology of the brain and body, in the quality of our attention and in the way we process emotion, interpret reality and meet pressure. Across this book, we have followed that thread closely. We have seen that leadership is not something we perform, but something that emerges from the condition of the human system beneath it.

We have watched what happens when that system is placed under continuous strain, and how modern life, with its speed, fragmentation and relentless demand for output, hollows out the very capacities leadership depends on. Under those conditions, even capable, intelligent and well-intentioned people lose contact with clarity, steadiness, empathy and discernment. Not because they lack knowledge, but because their system can no longer make that knowledge available when it matters most.

This is the premise that changes everything. It is also the premise that must now meet a world none of us have ever lived in before.

We are entering a time in which intelligence is no longer exclusively human. That shift is already underway, and it is not theoretical. Systems can analyse, generate, recommend and decide with a speed and sophistication that would have seemed impossible a decade ago. Work that once required hours of concentration can now be completed in seconds. In many ways, this is progress. It expands what we can reach, and it removes friction that once stood between intention and outcome.

But beneath that progress sits a question the human species has never had to ask itself with this much urgency. If intelligence is no longer what makes us distinct, then what is?

For a long time, we answered that question with output. We equated value with production, worth with achievement and contribution with visible result. Leadership was shaped by the same logic. The better leader was the one who could think faster, decide more quickly, execute more efficiently and deliver more impressively. That model is reaching the end of its usefulness. When intelligence becomes available on demand, performance alone can no longer define leadership. Speed alone can no longer define value. Output alone can no longer define contribution. The ground shifts from capability to meaning, from intelligence to wisdom, from performance to presence.

And this shift reaches further than most people realise. Because leadership, in the sense this book has been describing, is not a role. It is not a title, a position or a function assigned by an organisation. It is a state of the human system, and that state affects every person who comes near it, whether we have chosen to lead or not. A parent leads. A teacher leads. A partner leads. A friend leads. A colleague leads. A stranger on a difficult day can lead the mood of an entire room simply by how they carry themselves through it. Anyone whose inner condition shapes the inner condition of another human being is already leading, and that is all of us, all the time, in ways both large and quiet. This is why the work of becoming a brain-friendly leader does not belong to the few. It belongs to anyone willing to take responsibility for what they transmit.

To become a brain-friendly leader is to learn to work with your own biology rather than against it. It is to recognise that your brain is not an abstract thinking machine sitting above the rest of you, but a living organ woven into sleep, stress, movement, nutrition, emotion and attention.

It is to understand what happens to your thinking when you are exhausted, what happens to your patience when you are depleted and what happens to your judgement when you are rushed. It is to notice how your habits either support the person you are trying to become, or quietly erode them. And it is to develop a kind of awareness that does not retreat into reflection alone, but lives continuously alongside action, an awareness of what steadies you, what pulls you off centre, what repeatedly reveals itself in the moments you most want to meet with grace. This kind of awareness is not passive. It is the beginning of authorship, because what remains unseen cannot be regulated, and what cannot be regulated cannot be trusted under pressure.

At the practical level, what this asks of us is both simple and demanding. It asks for the capacity to remain clear in the middle of complexity, not because complexity disappears, but because the internal system learns to hold it. The conditions that support this are ordinary and available. Sleep that restores rather than depletes. Movement that generates energy rather than drains it. Breath that settles rather than accelerates. Moments of stillness that allow thought to deepen rather than collapse under speed. These are not luxuries reserved for quiet lives. They are the soil out of which coherent leadership grows. The aim is not perfection. It is alignment. And alignment, once it becomes familiar, begins to change everything it touches.

It changes everything it touches because no one leads in isolation. If you are chronically stressed, reactive or disconnected from yourself, that state does not stay inside you. It leaks into the room. It shapes the mood of a conversation, the atmosphere of a household, the feel of a team before a word is spoken. A dysregulated leader does not only experience fragmentation. They distribute it. A coherent leader does not only experience clarity. They make it available to others. Leadership, at this level, becomes a social force. Every interaction carries a signal. Every decision shapes an environment.

Over time, those signals accumulate into culture, family, community and the texture of the world we all live inside.

Becoming a brain-friendly leader is not a luxury. It is not a nice addition for those drawn to wellbeing or personal growth. It is a responsibility. A responsibility to understand yourself well enough not to harm others through patterns you have never examined. A responsibility to care for the brain and body that carry you through every relationship, every decision and every hour of every day. A responsibility to create conditions in which other people can think, feel and contribute without bracing against you. And ultimately, a responsibility to remain human in a world increasingly shaped by systems that are not.

The future will not be decided by what technology can do. It will be decided by what human beings choose to do in response. We can use these systems to accelerate, optimise and expand what we produce, or we can use them while also deepening what we are. We can become faster, or we can become wiser. These choices are not mutually exclusive, but neither do they happen on their own. The real challenge of this moment is not technological. It is developmental. Can we evolve our humanity fast enough to remain worthy of the systems we are building? That question will not be answered by governments or institutions. It will be answered by individuals, one by one, in the private decisions they make about how they treat their own minds.

This book has not been an argument against progress. It has been an argument for alignment, for making sure that as the external world grows more powerful, the internal world is not abandoned. It has been an argument for recognising that leadership does not begin with what we do, but with how we are, and that the most advanced systems ever created will not compensate for a human being whose own system is fragmented.

It has also been an argument for something more radical than it may appear on the surface: that the brain you live inside is not only yours to use, but yours to shape. That your attention is not a resource to be mined by whatever catches it, but a capacity to be guarded, directed and put in service of what you care about. That your nervous system is not a private matter, but the instrument through which you touch every person you encounter.

To take your brain into your own hands, in this sense, is not self-improvement. It is civic participation. It is the foundation of every other contribution you might hope to make. A more peaceful home begins with a more regulated nervous system. A more honest conversation begins with a steadier internal state. A more humane workplace begins with leaders who can stay present under pressure. A more thoughtful world begins with minds that have been given the conditions to think. None of this requires permission. None of it requires a title. None of it waits for someone else to begin. The work starts wherever you are, with whatever is in front of you, in whatever hour of whatever day you decide it matters.

And it does matter. It matters more now than at any point most of us have lived through. Because the world is accelerating in every direction at once, and the only counterweight available, the only thing that can hold the centre while everything else moves, is the quality of the human beings inside it. Not their credentials. Not their access. Not their intelligence measured against machines that will always calculate faster. What will hold is their coherence. Their presence. Their ability to remain themselves when the conditions around them are pulling in every other direction.

The world does not need more information, more noise or more speed for their own sake. It needs people who can stay present when the culture rewards distraction. Who can think deeply when the default is reaction. Who can feel without being swept away. Who can act without losing themselves in the act.

People who understand that what they are is always shaping what they do, and that what they do is always shaping someone else.

That is the work. And it begins closer than we tend to think, in the brain, in the body, in the patterns we repeat without noticing, and in the choice, available in every single moment, to lead from coherence rather than reaction.

If we want a better world, we will not build it through behaviour alone. We will build it through the quality of the human systems behind that behaviour. We will build it through people who have decided, personally, that the state of their own mind is worth protecting, not only for themselves but for everyone their life touches.

Because in the end, the future of leadership is not something that will happen to us. It is something we are already shaping, breath by breath, decision by decision, moment by moment, whether we realise it or not.

The only question left is what kind of human we are willing to become while the world is changing around us.

And that question belongs to all of us.

“

**WHAT WILL ENDURE
IS NOT INTELLIGENCE ALONE,
BUT THE HUMANITY THAT LEARNS
HOW TO LIVE WITH IT.**

”

THE MEANING BEHIND THE COVER

The cover represents a moment of transition, where the human stands between what has been and what is now unfolding. The dark silhouette reflects the unknown, not as absence, but as the space where growth and transformation take shape.

Within it, the brain, heart, and gut symbolise the integration of thinking, feeling, and instinct, a reminder that leadership emerges from the whole human system rather than from any single part of it. As the figure dissolves into pixels, it points to a shift from fixed ways of being toward greater adaptability, awareness, and responsiveness in an increasingly digital world.

The surrounding gradient suggests an expansion beyond traditional boundaries, where leadership is no longer confined to structure or role, but becomes more fluid, connected, and dynamic. This is not about losing our humanity, but about evolving it, learning to navigate complexity while staying grounded in the intelligence that already exists within us.

ABOUT THE AUTHORS

SILVIA DAMIANO

Silvia Damiano is a scientist, author, speaker, filmmaker and founder of the About my Brain Institute. For more than twenty-five years, she has explored one of the defining questions of our time: how do we remain deeply human in a world that is becoming faster, more distracted and increasingly shaped by artificial intelligence?

Born in Argentina and based in Australia, Silvia is the creator of the i4 Neuroleader Model, a pioneering framework that integrates neurobiology, emotional intelligence, and creativity to strengthen the human capacities most needed for the future. Her work emerged not only from years of formal research and global leadership development, but also from a deeply personal inquiry into what happens when the human system is pushed beyond its limits, and what it takes to restore coherence, wellbeing, and inner strength.

Through her books, global programs, and public speaking, Silvia has become a compelling voice on the future of leadership, human connection, and the inner conditions that allow people to think clearly, live meaningfully, and stay whole under pressure. She is the author of Leadership Is Upside Down, co-author of Brain-Friendly Leadership: The 9 Habits of a Future-Fit Mind, and the producer and director of the award-winning documentary Make Me A Leader. Her work translates complex science into a deeply human message: that in an age of acceleration, distraction, and AI, our greatest responsibility is not only to perform, but to protect what makes us human.

In Brain-Friendly Leader, Silvia invites readers into a new leadership premise, one that begins not with behaviour, but with the condition of the human system beneath it. Her work is a call to reclaim presence, emotional clarity, meaningful connection, and the capacity to lead without losing ourselves in the process.

JUAN CARLOS CUBEIRO

Juan Carlos Cubeiro is a leading international expert in leadership, talent, and transformation, with more than three decades of experience advising organisations and senior executives across Europe. Originally from Spain, he is widely recognised as one of the most influential voices in leadership in the Spanish-speaking world, and a recipient of the Premio Nacional de Management, with a strong reputation for shaping the future of talent and organisational performance.

He has led strategic consulting and transformation projects for over a thousand organisations, including many of the world's leading companies, and has coached more than 700 CEOs and senior leaders. His work spans business, sport, and society, helping individuals and teams navigate complexity, unlock potential, and achieve sustainable results in an ever-changing environment. Juan Carlos is the European President of the About my Brain Institute and co-author of Leadership Is Upside Down and Brain-Friendly Leadership: The 9 Habits of a Future-Fit Mind, contributing to the advancement of a neuroscience-based, human-centred approach to leadership. He is also Strategy and Leadership Director at Amrop, President of the Academic Council at EBS Business School, and a long-standing professor at leading institutions, where he bridges academic insight with real-world application.

A prolific author of more than sixty books and hundreds of articles, Juan Carlos is a passionate advocate of learnability, dedicating a significant part of his time to continuous learning and development. His thought leadership has introduced concepts such as talentism, emphasising that talent, more than capital, is the defining factor of success in the modern economy.

Recognised internationally for his contribution to leadership, education, and human development, Juan Carlos continues to inspire leaders across sectors to think differently, grow continuously, and place people at the centre of transformation.

RELMI DAMIANO

Relmi Damiano is the Co-Founder and Creative Director of the About my Brain Institute, where she has played a pivotal role in shaping the global movement toward brain-friendly leadership since its inception in 2009. Blending creativity, design, and personal development, her work brings to life innovative learning experiences that empower individuals and organisations to grow, transform, and thrive.

With a background as an award-winning artist and photographer, Relmi began her career in the creative industries before expanding into digital media, branding, and experience design. Her work has been exhibited across Australia and internationally, and she has collaborated with clients across Europe, the United States, and Australia.

At the Institute, Relmi leads the creative vision and development of products, programs, and immersive experiences that have supported the certification of thousands of practitioners worldwide in the i4 Neuroleader™ Methodology. She also served as art director and producer of the award-winning documentary Make Me A Leader, and has spoken publicly, including her TEDx talk Biohacking Your Creative Self, in which she explores the distinction between imagination and creativity and the practical disciplines that allow both to flourish.

Alongside her creative leadership, Relmi has spent over a decade as a lecturer and mentor at institutions including Billy Blue College of Design and The University of Sydney, teaching subjects such as design innovation, entrepreneurship, branding and user experience.

Her work is further enriched by a deep commitment to contemplative practice and intuitive development. She is a certified yoga and meditation teacher, an advanced sound healing practitioner, and an intuitive coach, guiding individuals in accessing deeper levels of awareness, creativity and alignment with their purpose.

LEARNING TOOLS & REFERENCES

OTHER BOOKS

Our books are designed to expand how you understand leadership and the human system. Grounded in the i4 Neuroleader™ Model, they translate neuroscience into clear, practical insight you can apply in real situations. Each chapter invites reflection on how you think, feel and act, helping you build awareness and develop habits that support more intentional, consistent leadership.

WWW.ABOUTMYBRAIN.COM/BOOKS

LEADERSHIP TRAINING

Our leadership training programs bring the i4 Neuroleader™ Model to life through practical application. Designed for individuals, teams and organisations, they focus on integrating the brain and body to strengthen performance, collaboration, innovation and agility. The experience combines science, reflection, real-world practice and content for the award-winning Make Me A Leader Documentary, to help leaders build capability that holds under pressure, not just in theory.

WWW.ABOUTMYBRAIN.COM/TRAINING

ASSESSMENT

The i4 Neuroleader™ Assessment is a neuroscience-based self and 360 feedback tool designed to measure and develop leadership at a deeper level. It evaluates how you operate across performance, collaboration, innovation and agility through sixteen underlying pillars. By combining your own perspective with feedback from others, it provides a holistic view of how your leadership is experienced in real life. It offers a clear and practical roadmap to grow your leadership and optimise how your brain and body perform together.

WWW.ABOUTMYBRAIN.COM/ASSESSMENT

CARDS

The Heart 2 Heart Conversation Cards are designed to unlock the conversations most people avoid but need the most. Each card is a structured prompt grounded in the i4 Neuroleader™ Model, created to disrupt habitual thinking and activate deeper reflection across the four leadership dimensions. Used in teams, coaching or personal development, they create space for honest dialogue, stronger connection and the kind of awareness that turns insight into meaningful action.

WWW.ABOUTMYBRAIN.COM/CARDS

CHAPTER 1

Arnsten, Amy F. T. "Stress Signalling Pathways That Impair Prefrontal Cortex Structure and Function." Nature Reviews Neuroscience 10, no. 6 (2009).

Craig, A. D. "How Do You Feel? Interoception: The Sense of the Physiological Condition of the Body." Nature Reviews Neuroscience 3, no. 8 (2002).

Damasio, Antonio. Descartes' Error: Emotion, Reason, and the Human Brain. Putnam, 1994.

Porges, Stephen W. The Polyvagal Theory: Neurophysiological Foundations of Emotions, Attachment, Communication, and Self-Regulation. Norton, 2011.

Siegel, Daniel J. The Developing Mind: How Relationships and the Brain Interact to Shape Who We Are. 2nd ed. Guilford Press, 2012.

Walker, Matthew. Why We Sleep: Unlocking the Power of Sleep and Dreams. Scribner, 2017.

CHAPTER 2

Adee, Sally. We Are Electric: The New Science of Our Body's Electrome. Canongate, 2023.

Barrett, Lisa Feldman, and W. Kyle Simmons. "Interoceptive Predictions in the Brain." Nature Reviews Neuroscience 16, no. 7 (2015).

Levin, Michael. "Molecular Bioelectricity in Developmental Biology: New Tools and Recent Discoveries." BioEssays 34, no. 3 (2012).

CHAPTER 3

Fisher, Jonathan. Just One Heart: A Cardiologist's Guide to Healing, Health, and Happiness. Manuscripts, 2024.

Glaser, Judith E. Conversational Intelligence: How Great Leaders Build Trust and Get Extraordinary Results. Bibliomotion, 2014.

Mayer, Emeran A. The Mind-Gut Connection: How the Hidden Conversation Within Our Bodies Impacts Our Mood, Our Choices, and Our Overall Health. Harper Wave, 2016.

McCraty, Rollin, and Doc Childre. "Coherence: Bridging Personal, Social, and Global Health." Alternative Therapies in Health and Medicine 16, no. 4 (2010).

Ram, Kaushik. "Danger of Replacing Instinct with Technology." TEDx talk, November 2024.

Soosalu, Grant, and Marvin Oka. mBraining: Using Your Multiple Brains to Do Cool Stuff. CreateSpace Independent Publishing Platform, 2012.

Thayer, Julian F., and Richard D. Lane. "Claude Bernard and the Heart-Brain Connection: Further Elaboration of a Model of Neurovisceral Integration." Neuroscience and Biobehavioral Reviews 33, no. 2 (2009).

CHAPTER 4

Benson, Herbert. The Relaxation Response. William Morrow, 1975.

Hatfield, Elaine, John T. Cacioppo, and Richard L. Rapson. Emotional Contagion. Cambridge University Press, 1994.

Langer, Ellen J. Mindfulness. Addison-Wesley, 1989.

CHAPTER 5

Arnsten, Amy F. T. "Stress Signalling Pathways That Impair Prefrontal Cortex Structure and Function." Nature Reviews Neuroscience 10, no. 6 (2009).

Brackett, Marc. Permission to Feel: The Power of Emotional Intelligence to Achieve Well-Being and Success. Celadon Books, 2019.

Edmondson, Amy. "Psychological Safety and Learning Behavior in Work Teams." Administrative Science Quarterly 44, no. 4 (1999).

Maté, Gabor. When the Body Says No: The Cost of Hidden Stress. Scribe Publications, 2019.

McEwen, Bruce S. "Protective and Damaging Effects of Stress Mediators." The New England Journal of Medicine 338, no. 3 (1998).

Porges, Stephen W. The Polyvagal Theory: Neurophysiological Foundations of Emotions, Attachment, Communication, and Self-Regulation. W. W. Norton & Company, 2011.

van der Kolk, Bessel A. The Body Keeps the Score: Brain, Mind, and Body in the Healing of Trauma. Viking, 2014.

CHAPTER 6

Carr, Nicholas. The Shallows: What the Internet Is Doing to Our Brains. W. W. Norton & Company, 2010.

Mark, Gloria. Attention Span: A Groundbreaking Way to Restore Balance, Happiness and Productivity. Hanover Square Press, 2023.

Turkle, Sherry. Alone Together: Why We Expect More from Technology and Less from Each Other. Basic Books, 2011.

CHAPTER 8

Haidt, Jonathan. The Anxious Generation: How the Great Rewiring of Childhood Is Causing an Epidemic of Mental Illness. Penguin Press, 2024.

UNESCO. Recommendation on the Ethics of Artificial Intelligence. UNESCO, 2021.

CHAPTER 9

Chamorro-Premuzic, Tomas. Why Do So Many Incompetent Men Become Leaders? (And How to Fix It). Harvard Business Review Press, 2019.

Dweck, Carol S. Mindset: The New Psychology of Success. Random House, 2006.

Rock, David. "SCARF: A Brain-Based Model for Collaborating with and Influencing Others." Neuroleader™ship Journal 1 (2008).

CHAPTER 10

Castellanos, Nazareth. El puente donde habitan las mariposas: Biosofía de la respiración. Siruela, 2025.

D'Esposito, Mark, and Bradley R. Postle. "The Cognitive Neuroscience of Working Memory." Annual Review of Psychology 66 (2015).

Kotler, Steven, and Jamie Wheal. Stealing Fire: How Silicon Valley, the Navy SEALs, and Maverick Scientists Are Revolutionizing the Way We Live and Work. Dey St., 2017.

Rosenthal, Norman E. Super Mind: How to Boost Performance and Live a Richer and Happier Life Through Transcendental Meditation. TarcherPerigee, 2016.

Harung, Harald S., and Frederick Travis. Excellence through Mind-Brain Development: The Secrets of World-Class Performers. Farnham, UK: Ashgate, 2015.

CHAPTER 11

Keysers, Christian, and Valeria Gazzola. "Dissociating the Ability and Propensity for Empathy." Trends in Cognitive Sciences 18, no. 4 (2014).

Lansing, Alfred. Endurance: Shackleton's Incredible Voyage. McGraw-Hill, 1959.

Nadella, Satya. Hit Refresh: The Quest to Rediscover Microsoft's Soul and Imagine a Better Future for Everyone. HarperBusiness, 2017.

Wicker, Bruno, Christian Keysers, Julie Plailly, Jean-Pierre Royet, Vittorio Gallese, and Giacomo Rizzolatti. "Both of Us Disgusted in My Insula: The Common Neural Basis of Seeing and Feeling Disgust." Neuron 40, no. 3 (2003).

CHAPTER 12

Beaty, Roger E., Yoed N. Kenett, Alexander P. Christensen, Michael D. Rosenberg, Mathias Benedek, Qunlin Chen, Andreas Fink, Jiang Qiu, Thomas R. Kwapil, Michael J. Kane, and Paul J. Silvia. "Robust Prediction of Individual Creative Ability from Brain Functional Connectivity." Proceedings of the National Academy of Sciences 115, no. 5 (2018).

Brown, Stuart. Play: How It Shapes the Brain, Opens the Imagination, and Invigorates the Soul. Avery, 2009.

Buckner, Randy L., and Daniel C. Carroll. "Self-Projection and the Brain." Trends in Cognitive Sciences 11, no. 2 (2007).

Hassabis, Demis, and Eleanor A. Maguire. "Deconstructing Episodic Memory with Construction." Trends in Cognitive Sciences 11, no. 7 (2007).

Sahlberg, Pasi. Finnish Lessons 3.0: What Can the World Learn from Educational Change in Finland? 3rd ed. Teachers College Press, 2021.

CHAPTER 13

Cryan, John F., and Timothy G. Dinan. "Mind-Altering Microorganisms: The Impact of the Gut Microbiota on Brain and Behaviour." Nature Reviews Neuroscience 13, no. 10 (2012).

Damasio, Antonio R. Descartes' Error: Emotion, Reason, and the Human Brain. Putnam, 1994.

Klein, Gary. The Power of Intuition: How to Use Your Gut Feelings to Make Better Decisions at Work. Currency, 2004.

Kuo, Ming. "How Might Contact with Nature Promote Human Health?" Frontiers in Psychology 6 (2015).

Leong, David. The Neurobiology of Entrepreneurial Intuition: A Three-Brain Framework. Preprint, 2025.

Mayer, Emeran A. The Mind-Gut Connection: How the Hidden Conversation Within Our Bodies Impacts Our Mood, Our Choices, and Our Overall Health. New York: Harper Wave, 2016.

Park, Bum-Jin, Yoshifumi Tsunetsugu, Takahide Kasetani, Takahiro Kagawa, and Yoshifumi Miyazaki. "The Physiological Effects of Shinrin-yoku." Environmental Health and Preventive Medicine 15, no. 1 (2010).

CHAPTER 14

Han, Byung-Chul. Psychopolitics: Neoliberalism and New Technologies of Power. Translated by Erik Butler. Verso, 2017.

Ienca, Marcello, and Roberto Andorno. "Towards New Human Rights in the Age of Neuroscience and Neurotechnology." Life Sciences, Society and Policy 13, no. 1 (2017).

Zuboff, Shoshana. The Age of Surveillance Capitalism: The Fight for a Human Future at the New Frontier of Power. PublicAffairs, 2019.

CHAPTER 15

Acemoglu, Daron, and Simon Johnson. Power and Progress: Our Thousand-Year Struggle Over Technology and Prosperity. PublicAffairs, 2023.

Danaher, John. Automation and Utopia: Human Flourishing in a World Without Work. Harvard University Press, 2019.

Susskind, Daniel. A World Without Work: Technology, Automation, and How We Should Respond. Metropolitan Books, 2020.

CHAPTER 16

Winthrop, Rebecca, and Adam Barton. Leapfrogging Inequality: Remaking Education to Help Young People Thrive. Brookings Institution Press, 2018.

www.ingramcontent.com/pod-product-compliance
Lightning Source LLC
LaVergne TN
LVHW100312210926
848480LV00003B/1100

* 9 7 8 0 6 4 5 9 5 3 7 5 6 *